Statistics For Big Data

FOR DUMMIES®
A Wiley Brand

by Alan Anderson, PhD
with David Semmelroth

FOR DUMMIES®
A Wiley Brand

Contents at a Glance

Table of Contents

Introduction

*W*elcome to *Statistics For Big Data For Dummies!* Every day, what has come to be known as *big data* is making its influence felt in our lives. Some of the most useful innovations of the past 20 years have been made possible by the advent of massive data-gathering capabilities combined with rapidly improving computer technology.

For example, of course, we have become accustomed to finding almost any information we need through the Internet. You can locate nearly anything under the sun immediately by using a search engine such as Google or DuckDuckGo. Finding information this way has become so commonplace that Google has slowly become a verb, as in "I don't know where to find that restaurant — I'll just Google it." Just think how much more efficient our lives have become as a result of search engines. But how does Google work? Google couldn't exist without the ability to process massive quantities of information at an extremely rapid speed, and its software has to be extremely efficient.

Another area that has changed our lives forever is e-commerce, of which the classic example is Amazon.com. People can buy virtually every product they use in their daily lives online (and have it delivered promptly, too). Often online prices are lower than in traditional "brick-and-mortar" stores, and the range of choices is wider. Online shopping also lets people find the best available items at the lowest possible prices.

Another huge advantage to online shopping is the ability of the sellers to provide reviews of products and recommendations for future purchases. Reviews from other shoppers can give extremely important information that isn't available from a simple product description provided by manufacturers. And recommendations for future purchases are a great way for consumers to find new products that they might not otherwise have known about. Recommendations are enabled by one application of big data — the use of highly sophisticated programs that analyze shopping data and identify items that tend to be purchased by the same consumers.

Although online shopping is now second nature for many consumers, the reality is that e-commerce has only come into its own in the last 15–20 years, largely thanks to the rise of big data. A website such as Amazon.com must process quantities of information that would have been unthinkably gigantic just a few years ago, and that processing must be done quickly and efficiently. Thanks to rapidly improving technology, many traditional retailers now also offer the option of making purchases online; failure to do so would put a retailer at a huge competitive disadvantage.

In addition to search engines and e-commerce, big data is making a major impact in a surprising number of other areas that affect our daily lives:

- Social media
- Online auction sites
- Insurance
- Healthcare
- Energy
- Political polling
- Weather forecasting
- Education
- Travel
- Finance

About This Book

This book is intended as an overview of the field of big data, with a focus on the statistical methods used. It also provides a look at several key applications of big data. Big data is a broad topic; it includes quantitative subjects such as math, statistics, computer science, and data science. Big data also covers many applications, such as weather forecasting, financial modeling, political polling methods, and so forth.

Our intentions for this book specifically include the following:

- Provide an overview of the field of big data.
- Introduce many useful applications of big data.
- Show how data may be organized and checked for bad or missing information.
- Show how to handle outliers in a dataset.
- Explain how to identify assumptions that are made when analyzing data.
- Provide a detailed explanation of how data may be analyzed with graphical techniques.

- Cover several key *univariate* (involving only one variable) statistical techniques for analyzing data.

- Explain widely used *multivariate* (involving more than one variable) statistical techniques.

- Provide an overview of modeling techniques such as regression analysis.

- Explain the techniques that are commonly used to analyze time series data.

- Cover techniques used to forecast the future values of a dataset.

- Provide a brief overview of software packages and how they can be used to analyze statistical data.

Because this is a *For Dummies* book, the chapters are written so you can pick and choose whichever topics that interest you the most and dive right in. There's no need to read the chapters in sequential order, although you certainly could. We do suggest, though, that you make sure you're comfortable with the ideas developed in Chapters 4 and 5 before proceeding to the later chapters in the book. Each chapter also contains several tips, reminders, and other tidbits, and in several cases there are links to websites you can use to further pursue the subject. There's also an online Cheat Sheet that includes a summary of key equations for ease of reference.

As mentioned, this is a big topic and a fairly new field. Space constraints make possible only an introduction to the statistical concepts that underlie big data. But we hope it is enough to get you started in the right direction.

Foolish Assumptions

We make some assumptions about you, the reader. Hopefully, one of the following descriptions fits you:

- You've heard about big data and would like to learn more about it.

- You'd like to use big data in an application but don't have sufficient background in statistical modeling.

- You don't know how to implement statistical models in a software package.

Possibly all of these are true. This book should give you a good starting point for advancing your interest in this field. Clearly, you are already motivated.

This book does not assume any particularly advanced knowledge of mathematics and statistics. The ideas are developed from fairly mundane mathematical operations. But it may, in many places, require you to take a deep breath and not get intimidated by the formulas.

Icons Used in This Book

Throughout the book, we include several icons designed to point out specific kinds of information. Keep an eye out for them:

A Tip points out especially helpful or practical information about a topic. It may be hard-won advice on the best way to do something or a useful insight that may not have been obvious at first glance.

A Warning is used when information must be treated carefully. These icons point out potential problems or trouble you may encounter. They also highlight mistaken assumptions that could lead to difficulties.

Technical Stuff points out stuff that may be interesting if you're really curious about something, but which is not essential. You can safely skip these if you're in a hurry or just looking for the basics.

Remember is used to indicate stuff that may have been previously encountered in the book or that you will do well to stash somewhere in your memory for future benefit.

Beyond the Book

Besides the pages or pixels you're presently perusing, this book comes with even more goodies online. You can check out the Cheat Sheet at www.dummies.com/cheatsheet/statisticsforbigdata.

We've also written some additional material that wouldn't quite fit in the book. If this book were a DVD, these would be on the Bonus Content disc. This handful of extra articles on various mini-topics related to big data is available at www.dummies.com/extras/statisticsforbigdata.

Where to Go From Here

You can approach this book from several different angles. You can, of course, start with Chapter 1 and read straight through to the end. But you may not have time for that, or maybe you are already familiar with some of the basics. We suggest checking out the table of contents to see a map of what's covered in the book and then flipping to any particular chapter that catches your eye. Or if you've got a specific big data issue or topic you're burning to know more about, try looking it up in the index.

Once you're done with the book, you can further your big data adventure (where else?) on the Internet. Instructional videos are available on websites such as YouTube. Online courses, many of them free, are also becoming available. Some are produced by private companies such as Coursera; others are offered by major universities such as Yale and M.I.T. Of course, many new books are being written in the field of big data due to its increasing importance.

If you're even more ambitious, you will find specialized courses at the college undergraduate and graduate levels in subject areas such as statistics, computer science, information technology, and so forth. In order to satisfy the expected future demand for big data specialists, several schools are now offering a concentration or a full degree in Data Science.

The resources are there; you should be able to take yourself as far as you want to go in the field of big data. Good luck!

Part I

Introducing
Big Data Statistics

getting started
with

**Big Data
Statistics**

In this part . . .

- Introducing big data and stuff it's used for
- Exploring the three Vs of big data
- Checking out the hot big data applications
- Discovering probabilities and other basic statistical idea

Chapter 1

What Is Big Data and What Do You Do with It?

In This Chapter

▶ Understanding what big data is all about

▶ Seeing how data may be analyzed using Exploratory Data Analysis (EDA)

▶ Gaining insight into some of the key statistical techniques used to analyze big data

Big data refers to sets of data that are far too massive to be handled with traditional hardware. Big data is also problematic for software such as database systems, statistical packages, and so forth. In recent years, data-gathering capabilities have experienced explosive growth, so that storing and analyzing the resulting data has become progressively more challenging.

Many fields have been affected by the increasing availability of data, including finance, marketing, and e-commerce. Big data has also revolutionized more traditional fields such as law and medicine. Of course, big data is gathered on a massive scale by search engines such as Google and social media sites such as Facebook. These developments have led to the evolution of an entirely new profession: the *data scientist,* someone who can combine the fields of statistics, math, computer science, and engineering with knowledge of a specific application.

This chapter introduces several key concepts that are discussed throughout the book. These include the characteristics of big data, applications of big data, key statistical tools for analyzing big data, and forecasting techniques.

Characteristics of Big Data

The three factors that distinguish big data from other types of data are *volume, velocity,* and *variety.*

Clearly, with big data, the *volume* is massive. In fact, new terminology must be used to describe the size of these datasets. For example, one *petabyte* of data consists of 1.0×10^{15} bytes of data. That's 1,000 *trillion* bytes!

A *byte* is a single unit of storage in a computer's memory. A byte is used to represent a single number, character, or symbol. A byte consists of eight *bits,* each consisting of either a 0 or a 1.

Velocity refers to the speed at which data is gathered. Big datasets consist of data that's continuously gathered at very high speeds. For example, it has been estimated that Twitter users generate more than a quarter of a million tweets *every minute.* This requires a massive amount of storage space as well as real-time processing of the data.

Variety refers to the fact that the contents of a big dataset may consist of a number of different formats, including spreadsheets, videos, music clips, email messages, and so on. Storing a huge quantity of these incompatible types is one of the major challenges of big data.

Chapter 2 covers these characteristics in more detail.

Exploratory Data Analysis (EDA)

Before you apply statistical techniques to a dataset, it's important to examine the data to understand its basic properties. You can use a series of techniques that are collectively known as *Exploratory Data Analysis* (EDA) to analyze a dataset. EDA helps ensure that you choose the correct statistical techniques to analyze and forecast the data. The two basic types of EDA techniques are *graphical* techniques and *quantitative* techniques.

Graphical EDA techniques

Graphical EDA techniques show the key properties of a dataset in a convenient format. It's often easier to understand the properties of a variable and the relationships between variables by looking at graphs rather than looking at the raw data. You can use several graphical techniques, depending on the type of data being analyzed. Chapters 11 and 12 explain how to create and use the following:

- ✔ Box plots
- ✔ Histograms
- ✔ Normal probability plots
- ✔ Scatter plots

Quantitative EDA techniques

Quantitative EDA techniques provide a more rigorous method of determining the key properties of a dataset. Two of the most important of these techniques are

- Interval estimation (discussed in Chapter 11).
- Hypothesis testing (introduced in Chapter 5).

Interval estimates are used to create a *range* of values within which a variable is likely to fall. *Hypothesis* testing is used to test various propositions about a dataset, such as

- The mean value of the dataset.
- The standard deviation of the dataset.
- The probability distribution the dataset follows.

Hypothesis testing is a core technique in statistics and is used throughout the chapters in Part III of this book.

Statistical Analysis of Big Data

Gathering and storing massive quantities of data is a major challenge, but ultimately the biggest and most important challenge of big data is putting it to good use.

For example, a massive quantity of data can be helpful to a company's marketing research department only if it can identify the key drivers of the demand for the company's products. Political polling firms have access to massive amounts of demographic data about voters; this information must be analyzed intensively to find the key factors that can lead to a successful political campaign. A hedge fund can develop trading strategies from massive quantities of financial data by finding obscure patterns in the data that can be turned into profitable strategies.

Many statistical techniques can be used to analyze data to find useful patterns:

- Probability distributions are introduced in Chapter 4 and explored at greater length in Chapter 13.
- Regression analysis is the main topic of Chapter 15.
- Time series analysis is the primary focus of Chapter 16.
- Forecasting techniques are discussed in Chapter 17.

Probability distributions

You use a *probability distribution* to compute the probabilities associated with the elements of a dataset. The following distributions are described and applied in this book:

- ✔ **Binomial distribution:** You would use the binomial distribution to analyze variables that can assume only one of two values. For example, you could determine the probability that a given percentage of members at a sports club are left-handed. See Chapter 4 for details.

- ✔ **Poisson distribution:** You would use the Poisson distribution to describe the likelihood of a given number of events occurring over an interval of time. For example, it could be used to describe the probability of a specified number of hits on a website over the coming hour. See Chapter 13 for details.

- ✔ **Normal distribution:** The normal distribution is the most widely used probability distribution in most disciplines, including economics, finance, marketing, biology, psychology, and many others. One of the characteristic features of the normal distribution is *symmetry* — the probability of a variable being a given distance below the mean of the distribution equals the probability of it being the same distance above the mean. For example, if the mean height of all men in the United States is 70 inches, and heights are normally distributed, a randomly chosen man is equally likely to be between 68 and 70 inches tall as he is to be between 70 and 72 inches tall. See Chapter 4 and the chapters in Parts III and IV for details.

  The normal distribution works well with many applications. For example, it's often used in the field of finance to describe the returns to financial assets. Due to its ease of interpretation and implementation, the normal distribution is sometimes used even when the assumption of normality is only approximately correct.

- ✔ **The Student's t-distribution:** The Student's t-distribution is similar to the normal distribution, but with the Student's t-distribution, extremely small or extremely large values are much more likely to occur. This distribution is often used in situations where a variable exhibits too much variation to be consistent with the normal distribution. This is true when the properties of small samples are being analyzed. With small samples, the variation among samples is likely to be quite considerable, so the normal distribution shouldn't be used to describe their properties. See Chapter 13 for details.

  Note: The Student's t-distribution was developed by W.S. Gosset while employed at the Guinness brewing company. He was attempting to describe the properties of small sample means.

✓ **The chi-square distribution:** The chi-square distribution is appropriate for several types of applications. For example, you can use it to determine whether a population follows a particular probability distribution. You can also use it to test whether the variance of a population equals a specified value, and to test for the independence of two datasets. See Chapter 13 for details.

✓ **The F-distribution:** The F-distribution is derived from the chi-square distribution. You use it to test whether the variances of two populations equal each other. The F-distribution is also useful in applications such as regression analysis (covered next). See Chapter 14 for details.

Regression analysis

Regression analysis is used to estimate the strength and direction of the relationship between variables that are *linearly* related to each other. Chapter 15 discusses this topic at length.

Two variables X and Y are said to be *linearly* related if the relationship between them can be written in the form

$$Y = mX + b$$

where

> m is the *slope,* or the change in Y due to a given change in X
>
> b is the *intercept,* or the value of Y when $X = 0$

As an example of regression analysis, suppose a corporation wants to determine whether its advertising expenditures are actually increasing profits, and if so, by how much. The corporation gathers data on advertising and profits for the past 20 years and uses this data to estimate the following equation:

$$Y = 50 + 0.25X$$

where

> Y represents the annual profits of the corporation (in millions of dollars).
>
> X represents the annual advertising expenditures of the corporation (in millions of dollars).

In this equation, the slope equals 0.25, and the intercept equals 50. Because the slope of the regression line is 0.25, this indicates that on average, for every $1 million increase in advertising expenditures, profits rise by $.25 million, or $250,000. Because the intercept is 50, this indicates that with no advertising, profits would still be $50 million.

This equation, therefore, can be used to forecast future profits based on planned advertising expenditures. For example, if the corporation plans on spending $10 million on advertising next year, its expected profits will be as follows:

$$Y = 50 + 0.25X$$
$$Y = 50 + 0.25(10) = 50 + 2.5 = 52.5$$

Hence, with an advertising budget of $10 million next year, profits are expected to be $52.5 million.

Time series analysis

A *time series* is a set of observations of a single variable collected over time. This topic is talked about at length in Chapter 16. The following are examples of time series:

- ✔ The daily price of Apple stock over the past ten years.
- ✔ The value of the Dow Jones Industrial Average at the end of each year for the past 20 years.
- ✔ The daily price of gold over the past six months.

With time series analysis, you can use the statistical properties of a time series to predict the future values of a variable. There are many types of models that may be developed to explain and predict the behavior of a time series.

One place where time series analysis is used frequently is on Wall Street. Some analysts attempt to forecast the future value of an asset price, such as a stock, based entirely on the history of that stock's price. This is known as *technical analysis*. Technical analysts do not attempt to use other variables to forecast a stock's price — the only information they use is the stock's own history.

Technical analysis can work only if there are inefficiencies in the market. Otherwise, all information about a stock's history should already be reflected in its price, making technical trading strategies unprofitable.

Forecasting techniques

Many different techniques have been designed to forecast the future value of a variable. Two of these are time series regression models (Chapter 16) and simulation models (Chapter 17).

Time series regression models

A *time series regression model* is used to estimate the trend followed by a variable over time, using regression techniques. A *trend line* shows the direction in which a variable is moving as time elapses.

As an example, Figure 1-1 shows a time series that represents the annual output of a gold mine (measured in thousands of ounces per year) since the mine opened ten years ago.

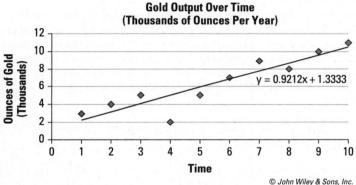

**Gold Output Over Time
(Thousands of Ounces Per Year)**

Figure 1-1:
A time series showing gold output per year for the past ten years.

© John Wiley & Sons, Inc.

The equation of the trend line is estimated to be

$$Y = 0.9212X + 1.3333$$

where

X is the year.

Y is the annual production of gold (measured in thousands of ounces).

This trend line is estimated using regression analysis. The trend line shows that on average, the output of the mine grows by 0.9212 thousand (921.2 ounces) each year.

You could use this trend line to predict the output next year (the 11th year of operation) by substituting 11 for X, as follows:

$$Y = 0.9212X + 1.3333$$
$$Y = 0.9212(11) + 1.3333 = 11.4665$$

Based on the trend line equation, the mine would be expected to produce 11,466.5 ounces of gold next year.

Simulation models

You can use *simulation* models to forecast a time series. Simulation models are extremely flexible but can be extremely time-consuming to implement. Their accuracy also depends on assumptions being made about the time series data's statistical properties.

Two standard approaches to forecasting financial time series with simulation models are historical simulation and Monte Carlo simulation.

Historical simulation

Historical simulation is a technique used to generate a probability distribution for a variable as it evolves over time, based on its past values. If the properties of the variable being simulated remain stable over time, this technique can be highly accurate. One drawback to this approach is that in order to get an accurate prediction, you need to have a lot of data. It also depends on the assumption that a variable's past behavior will continue into the future.

As an example, Figure 1-2 shows a histogram that represents the returns to a stock over the past 100 days.

This histogram shows the probability distribution of returns on the stock based on the past 100 trading days. The graph shows that the most frequent return over the past 100 days was a loss of 2 percent, the second most frequent was a loss of 3 percent, and so on. You can use the information contained within this graph to create a probability distribution for the most likely return on this stock over the coming trading day.

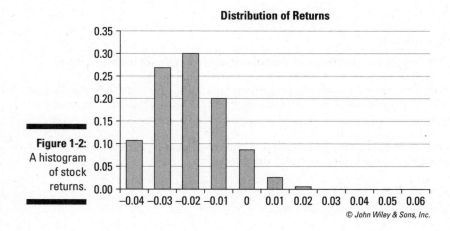

Figure 1-2: A histogram of stock returns.

© John Wiley & Sons, Inc.

Monte Carlo simulation

Monte Carlo simulation is a technique in which random numbers are substituted into a statistical model in order to forecast the future values of a variable. This methodology is used in many different disciplines, including finance, economics, and the hard sciences, such as physics. Monte Carlo simulation can work very well but can also be extremely time-consuming to implement. Also, its accuracy depends on the statistical model being used to describe the behavior of the time series.

As you can see, we've got a lot to cover in this book. But don't worry, we take it step by step. In Part I, we look at what big data is. We also build a statistical toolkit that we carry with us throughout the rest of the book. Part II focuses on the (extremely important) process of preparing data for the application of the techniques just described. Then we get to the good stuff in Parts III and IV. Though the equations can appear a little intimidating at times, we have labored to include examples in every chapter that make the ideas a little more accessible. So, take a deep breath and get ready to begin your exploration of big data!

Chapter 2

Characteristics of Big Data: The Three Vs

● ●

In This Chapter

▶ Understanding the characteristics of big data and how it can be classified

▶ Checking out the features of the latest methods for storing and analyzing big data

● ●

*T*he phrase *big data* refers to *datasets* (collections of data) that are too massive for traditional database management systems (DBMS) to handle properly. The rise of big data has occurred for several reasons, such as the massive increase in e-commerce, the explosion of social media usage, the advent of video and music websites, and so forth.

Big data requires more sophisticated approaches than those used in the past to handle surges of information. This chapter explores the characteristics of big data and introduces the newer approaches that have been developed to handle it.

Characteristics of Big Data

The three main characteristics that define big data are generally considered to be volume, velocity, and variety. These are the three Vs. *Volume* is easy to understand. There's *a lot* of data. *Velocity* suggests that the data comes in faster than ever and must be stored faster than ever. *Variety* refers to the wide variety of data structures that may need to be stored. The mixture of incompatible data formats provides another challenge that couldn't be easily managed by DBMS.

Volume

Volume refers, as you might expect, to the quantity of data being generated. A proliferation of new sources generates massive amounts of data on a continuous basis. The sources include, but are certainly not limited to, the following:

- Internet forums
- YouTube
- Facebook
- Twitter
- Cellphones (videos, photos, texts)
- Internet search engines
- Political polling

The volume of data being created is accelerating rapidly, requiring new terminology to describe these massive quantities. This terminology includes names that describe progressively larger amounts of storage. These names can sound quite strange in a world where people are familiar with only megabytes (MB) and gigabytes (GB), and maybe terabytes (TB). Some examples are the *petabyte* (PB), the *zettabyte* (ZB), and the *yottabyte* (YB).

You are likely familiar with the megabyte: one thousand kilobytes, or one million bytes of storage. A gigabyte refers to one *billion* bytes of storage. Until recently, the storage capacity of hard drives and other storage devices was in the range of hundreds of gigabytes, but in 2015 1TB, 2TB, and 4TB internal and external hard drives are now common.

The next step up is the terabyte, which refers to one *trillion* bytes. One trillion is a *large* number, expressed as a one followed by *twelve* zeros:

1,000,000,000,000

You can write this number using *scientific notation* as 1.0×10^{12}.

With scientific notation, a number is expressed as a constant multiplied by a power of ten. For example, 3,122 would be expressed as 3.122×10^3, because 10^3 equals 1,000. The constant always has one digit before the decimal point, and the remaining digits come after the decimal point.

For larger units of storage, the notation goes like this:

1.0×10^{15} bytes = one petabyte

1.0×10^{18} bytes = one exabyte

1.0×10^{21} bytes = one zettabyte

1.0×10^{24} bytes = one yottabyte

Here's an interesting name for a *very* large number: 1.0×10^{100} is called a *googol*. The name of the search engine Google is derived from this word. Speaking of Google, the company is currently processing over 20 petabytes of information each day, which is more than the estimated amount of information currently stored at the Library of Congress.

Velocity

As the amount of available data has surged in recent years, the speed with which it becomes available has also accelerated dramatically. Rapidly received data can be classified as the following:

✔ Streaming data

✔ Complex event processing

Streaming data is data transferred to an application at an extremely high speed. The classic example would be the movies you download and watch from sources such as Netflix and Amazon. In these cases, the data is being downloaded while the movie is playing. If your Internet connection isn't very fast, you've probably noticed annoying interruptions or glitches as the data downloads. In those cases, you need more *velocity*.

Streaming is useful when you need to make decisions in real time. For example, traders must make split-second decisions as new market information becomes available. An entire branch of finance known as *market microstructure* analyzes how prices are generated based on real-time trading activity. *High-frequency trading* (HFT) uses computer algorithms to generate trades based on incoming market data. The data arrives at a high speed, and the assets are held for only fractions of a second before being resold.

Complex event processing (CEP) refers to the use of data to predict the occurrence of events based on a specific set of factors. With this type of processing, data is examined for patterns that couldn't be found with more traditional approaches, so that better decisions may be made in real time. An example is your GPS device's ability to reroute you based on traffic and accident data.

Variety

In addition to traditional data types (numeric and character fields in a file), data can assume a large number of different forms. Here are just a few:

- ✔ Spreadsheets
- ✔ Word-processing documents
- ✔ Videos
- ✔ Photos
- ✔ Music
- ✔ Emails
- ✔ Text messages

With such a variety of formats, storing and analyzing these kinds of data are extremely challenging. The formats are incompatible with each other, so combining them into one large database is problematic.

This is one of the major challenges of big data: finding ways to extract useful information from multiple types of disparate files.

Traditional Database Management Systems (DBMS)

A traditional DBMS stores data and enables it to be easily retrieved. There are several types of database management systems, which can be classified according to the way data is organized and cross-referenced. This section focuses on three of the most important types: relational model, hierarchical model, and network model databases.

Relational model databases

With a relational database, the data is organized into a series of *tables*. Data is accessed by the row and column in which it's located. This model is very flexible and is easy to expand to include new information. You simply add more records to the bottom of an existing table, and you can create new categories by simply adding new rows or columns.

Table 2-1 shows a simple example of a table in a relational database.

Table 2-1	Employee Data Organized as a Relational Database		
Name	Title	Years with Company	Annual Salary
Smith, John	Senior Accountant	8	$144,000
Jones, Mary	VP, Research and Development	24	$250,000
Williams, Tony	CFO	13	$210,000
. . .	. . .	. . .	. . .

The data in Table 2-1 is organized as a series of *records in a table*. Each record contains information about one employee. Each record contains four *fields:* Name, Title, Years with Company, and Annual Salary.

Using this setup, you can find information about employees very quickly and easily. For example, if the human resources department wants to determine which employees have been with the company for at least ten years, a new table — with information drawn from this table — could be generated to list the employees. Table 2-2 shows the new table.

Table 2-2	Employees Who Have Been with the Company at Least Ten Years
Name	Years with Company
Jones, Mary	24
Williams, Tony	13
. . .	. . .

The relational database user accesses the information with a special type of software known as a *query language*. One of the most widely used query languages is SQL (Structured Query Language).

The "structure" of Structured Query Language is quite simple and is basically the same for all relational database systems. Syntax differs slightly from system to system. But in all cases, queries follow the same format (though not all elements need always be present).

Select (list of data fields you want to see)

From (list of tables containing the data)

Where (list of filtering and other conditions you want to use)

Group by (instructions for summarizing the data)

Having (list of conditions on the summarized data)

Order by (sorting instructions).

So for example, the report shown in Table 2-2 could have been generated by the following query:

Select Name, Years with Company

From Employee Data

Where Years with Company >10.

Currently, the relational model is the most commonly used type of DBMS. Some examples of relational database software include the following:

- SQL Server
- Microsoft Access
- DB2
- MySQL
- Oracle

Hierarchical model databases

A hierarchical database is organized as a *tree*. Records are connected through links. This type of database has a top-down structure, with all searches for data starting at the top of the structure and continuing through the entire structure until the desired information is found.

For example, Figure 2-1 shows a diagram of a hierarchical database. The database contains student records at a university. The students are organized according to whether they attend the School of Business or the School of Arts and Sciences. Within each school, the students are further classified as undergraduates or graduate students.

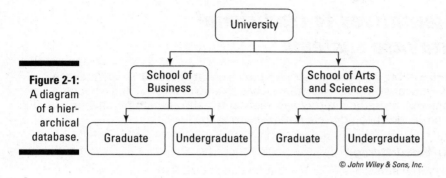

Figure 2-1:
A diagram of a hierarchical database.

You can think of each box in the diagram as a *node,* and each arrow as a *branch.* The University node is the *parent* of the School of Business and School of Arts and Sciences nodes. The School of Business node is a *child* of the University node, as is the School of Arts node.

One of the drawbacks of this model is that accessing a record requires searching through the entire structure until the record is found — and that's extremely time-consuming. For example, finding the record of a specific graduate business student requires starting with the University node, followed by the School of Business node, and then the Graduate node.

Another drawback to this model is that each parent node may have many child nodes, but each child node may only have one parent node. For many types of data, this doesn't accurately describe the relationship among the records.

Hierarchical models are not nearly as prevalent as relational systems. They are useful when the data you are managing actually is a hierarchy. Perhaps the most familiar such instances are file managers, such as the Finder on the Mac and Windows Explorer in Windows.

Network model databases

The network model is a more flexible version of the hierarchical model. It's also organized as a tree with branches and nodes. However, one important difference between the two models is that the network model allows for each child node to have more than one parent node. Because of this, much more complex relationships may be represented.

Again, these network models are not as widespread as the relational model. One place where they have been used extensively is in geographic information systems. The fact that road intersections have multiple branches makes the network model convenient.

Alternatives to traditional database systems

The rise of big data has outstripped the capacity of traditional database management systems. Two approaches to addressing this have become commonplace in the Internet age: distributed storage and parallel processing. The basic idea behind them both is sharing the load.

Distributed storage

Distributed storage is exactly what it sounds like. Rather than gather all the data into a central location, the data is spread out over multiple storage devices. This allows quicker access because you don't need to cull through a huge file to find the information you're looking for.

Distributed storage also allows for more frequent backups. Because systems are writing data to a lot of small files, real-time backups become reasonable.

Distributed storage is the backbone of so-called *cloud computing*. Many find it reassuring that all the books, music, and games they have ever purchased from the Web are backed up in the cloud. Even if you drop your iPad in the lake, for example, you could have everything restored and available on a new device with very little effort.

Parallel processing

Distributed storage allows another type of sharing to be done. Because the data is stored on different machines, it can be analyzed using different processors. This is known as *parallel processing*. It is particularly useful for mathematical and statistical applications involving very complex modeling techniques.

Even with the very powerful computers available today, big data analysis would still be impossible without parallel processing. The human genome project is wholly dependent on having a server farm to sort out the seemingly infinite number of possibilities.

Parallel processing can be very widely distributed. To illustrate, there is a climate prediction project that has been managed through Oxford University for a little over a decade. The website Climateprediction.net manages a distributed computing array that is borrowing resources from almost 30,000 machines. There are similar arrays searching for large prime numbers that number in the thousands.

Chapter 3

Using Big Data: The Hot Applications

In This Chapter

▶ Understanding the impact that big data is making in several diverse fields

▶ Checking out new products and services stemming from big data

▶ Considering the challenges facing users of big data

*T*hanks to the continuing surge of computing power and storage capacity, it has become possible to gather, process, and analyze quantities of data that would have been unimaginable just a decade ago. This has given rise to the entirely new field of big data. Big data bridges several disciplines, including statistics, mathematics, and computer science. It addresses the unique challenges associated with processing enormous volumes of information. Big data is already making major inroads into a wide variety of highly diversified fields, ranging from online shopping to healthcare services.

This chapter introduces several of the most exciting areas in which big data is having a major impact. In many cases, the acceleration of computer technology is increasing efficiency, lowering costs, making new services available, and improving the quality of life. Some of these areas include the following:

✔ Weather forecasting

✔ Healthcare

✔ Insurance

✔ Finance

✔ Electric utilities

✔ Higher education

✔ Retail services and online shopping

✔ Search engines

✔ Social media

Of these fields, online shopping and search engines couldn't exist at all without big data. Weather forecasting has benefited tremendously from the massive increase in data-processing speeds and data-gathering capabilities that has occurred in recent years. Other fields, such as retail services, finance, banking, insurance, education, and so forth, certainly predated the rise of big data, but have rapidly adopted it in order to

- Gain a competitive edge.
- Produce new types of products and services.
- Improve the quality of existing products and services.
- Lower production costs.

The rise of big data has also led to an increase in the demand for quantitative and programming skills — and is likely to generate a large number of high-paying jobs in the near future.

Big Data and Weather Forecasting

Weather forecasting has always been extremely challenging, given the number of variables involved and the complex interactions between those variables. Dramatic increases in the ability to gather and process data have greatly enhanced the ability of weather forecasters to pinpoint the timing and severity of hurricanes, floods, snowstorms, and other weather events.

One example of an application of big data to weather forecasting is IBM's Deep Thunder. Unlike many weather forecasting systems, which give general information about a broad geographical region, Deep Thunder provides forecasts for extremely specific locations, such as a single airport, so that local authorities can get critically important information in real time. Here are some examples of the information that Deep Thunder can provide:

- Estimates of areas where flooding is likely to be most severe
- The strength and direction of tropical storms
- The most likely amount of snow or rain that will fall in a specific area
- The most likely locations of downed power lines
- Estimates of areas where wind speeds are likely to be greatest
- The locations where bridges and roads most likely to be damaged by storms
- The likelihood of flights being cancelled at specific airports

This information is essential for emergency planning. Using big data, local authorities can better anticipate problems caused by weather before they occur. For example, planners can make preparations to evacuate low-lying areas that are likely to be flooded. It's also possible to make plans to upgrade existing facilities. (For example, power lines that are prone to being disabled by heavy winds can be upgraded.)

One important customer of Deep Thunder is the city of Rio de Janeiro, Brazil, which will be using the system in planning for the 2016 Olympics. Using the technology, the city will make use of improved forecasts for storms, floods, and other natural disasters in order to ensure that the Olympics won't be disrupted by such events.

IBM is also providing massive computing power to the Korean Meteorological Administration (KMA) to fully embrace big data technology. The KMA gathers over 1.5 terabytes of meteorological data each day, which requires a staggering amount of storage and processing power to analyze. By using big data, the KMA will be able to improve its forecasts regarding the strength and location of tropical storms and other weather systems.

A terabyte is equal to one trillion bytes. That's 1,000,000,000,000 bytes of information. You'd write one trillion bytes in scientific notation as 1.0×10^{12}. To put that in perspective, you would need around 1,500 CDs to store a single terabyte. Including their plastic cases, that would stack up as a 40-foot tall tower of CDs.

Another example of using big data in weather forecasting took place during Hurricane Sandy in 2012 — the "storm of the century." The National Hurricane Center was able to use big data technology to predict the hurricane's landfall to within 30 miles a full five days in advance. That is a dramatic increase in accuracy from what was possible even 20 years ago. As a result, FEMA and other disaster management organizations were far better prepared to deal with the mess than they might have been had it occurred in the 1990s or earlier.

One of the interesting consequences of gathering and processing more weather data is the appearance of corporations that sell customized insurance to protect against weather damage. One example is the Climate Corporation, which was formed in 2006 by two former employees of Google. The Climate Corporation sells weather-forecasting services and specialized insurance to farmers seeking to hedge the risk of crop damage. The company uses big data to pinpoint the types of risks that are relevant to a specific area, based on massive amounts of data on moisture, soil type, past crop yields, and so on.

Farming is an exceptionally risky business, because the variable of weather is far less predictable than the variables that affect most other businesses, such as interest rates, the state of the economy, and so forth. Although farm insurance is available from the federal government, in many cases it isn't sufficient to meet the more specialized types of risks that plague individual famers. The Climate Corporation fills gaps in federal insurance — gaps that would be impossible to offer without an improved understanding of the risk factors facing individual farmers. In the future, as more data becomes available, even more specialized insurance products (such as insurance for specific crops) may become available.

Big Data and Healthcare Services

Healthcare is one area where big data has the potential to make dramatic improvements in the quality of life. The increasing availability of massive amounts of data and rapidly increasing computer power could enable researchers to make breakthroughs, such as the following:

- Predicting outbreaks of diseases
- Gaining a better understanding of the effectiveness and side effects of drugs
- Developing customized treatments based on patient histories
- Reducing the cost of developing new treatments

One of the biggest challenges facing the use of big data in healthcare is that much of the data is stored in independent "silos." A *data silo* is a collection of data that isn't used on a regular basis and so isn't accessible outside of the silo. Healthcare data comes from multiple sources:

- Public health records
- Government databases
- Insurance companies
- Pharmaceutical companies
- Patient health histories
- Personal tracking devices

Much of the data is scattered and not readily accessible. Also, the data may be stored in many different formats — some of it still on paper! As a result, much information that could be potentially useful in some applications may be difficult and time-consuming to acquire.

Once this data has been combined properly, the potential exists to dramatically improve the analytical techniques used to diagnose patients. For example, it may eventually become possible to provide genetic sequencing for each individual patient; at the moment, this would be a prohibitively expensive and lengthy process in most cases.

Here are some other potential benefits of using big data in healthcare:

- ✔ Increased ability of patients to monitor their own treatments
- ✔ Improved ability of doctors to choose the best treatments for patients
- ✔ More efficient matching of patients with the appropriate healthcare professionals

Another potential benefit comes in the area of controlling costs. As the availability of healthcare data increases, the potential for reducing costs through better preventative treatment, increased efficiency of the drug development cycle, improved monitoring of patients, and other methods appears to be quite substantial. The consulting firm McKinsey & Company estimated in 2011 that the potential savings could be $300 billion per year — a number that could grow over time.

Big Data and Insurance

The insurance industry couldn't survive without the ability to gather and process substantial quantities of data. In order to determine the appropriate premiums for their policies, insurance companies must be able to analyze the risks that policyholders face and be able to determine the likelihood of these risks actually materializing.

Due to substantial increases in the availability of data and the speed and storage capacity of computers, new opportunities have arisen for the insurance industry to increase profits through the improved modeling of risks, the use of more efficient pricing practices, and the ability to offer more specialized products. Additionally, big data may be used for security purposes, such as detecting and preventing insurance fraud.

A good example of the use of big data in the insurance industry is the growing use of telematics devices in the auto insurance industry.

A *telematics* device transmits computer data wirelessly. It may be used for many purposes, such as increasing the quality of a product or ensuring the safety of a process. A Global Positioning System (GPS) is an example of a telematics device.

An auto insurance company may install a telematics device in a vehicle, with the resulting data transmitted to the insurance company in real time. The data may include the following details:

- Speed
- Number of miles driven
- Braking patterns
- Time of day when driving takes place

These data can help an insurance company determine the likelihood of a given driver becoming involved in an accident. The company can use this information to set the premium paid by the individual driver.

The benefit to the driver is that he or she may be eligible for lower premiums if the data show a pattern of safe driving. Another benefit is that the driver will have a better understanding of his or her own driving habits, and will gain knowledge about how to drive more safely.

One of the drawbacks of using telematics devices is the need to process and store significant amounts of data. Another potential issue is that insurance companies may receive telematics data from multiple providers, raising the possibility of data-compatibility issues.

Another huge challenge for insurance companies is identifying and quantifying the most important risk factors from the massive amounts of data being gathered. For example, the insurance company must decide how much each mile driven contributes to the likelihood of an accident. This requires a great deal of sophisticated statistical modeling.

Despite these potential problems, the use of telematics devices in the auto insurance industry is expected to grow rapidly in the next few years as the ability to process the required data continues to improve and public acceptance of the idea grows.

Telematics is currently being used far more widely in commercial auto insurance than for personal auto owners. Fleets of trucks and taxis are good examples. But it is beginning to move into the personal auto space on a voluntary basis. Everybody thinks they are a good driver and wants to get a discount for being one.

But this does raise a larger point about big data and privacy. With all this data floating around, where is the line drawn about what companies and governments can legally know about you? There is no simple answer to that question and in fact it is a constant topic of debate in Congress. Beyond your driving habits, everything from location tracking on your mobile device to which websites you surf is potentially out there to be had. And given people's apparent willingness to sacrifice privacy for convenience, it's worth keeping an eye on what companies are doing with your personal data.

The increased use of telematics devices may also provide additional benefits to society, as the data should make it possible for local authorities to improve the safety of roads and bridges by analyzing the factors that are most likely to contribute to accidents.

Big Data and Finance

One area of the finance industry that has been dramatically affected by big data is the trading activities of banks and other financial institutions. An example is high-frequency trading (HFT), a relatively new mode of trading that depends on the ability to execute massive volumes of trades in extremely short time intervals. HFT traders make money by executing a huge number of trades, each of which earns a miniscule profit. Unlike traditional traders, HFT traders don't attempt to hold positions for any great length of time and don't base their trades on fundamental factors such as interest rates, exchange rates, commodity prices, and so forth. The success of HFT trades depends critically on the speed of execution, as they are based on rapid fluctuations in market prices.

As more and more resources have been dedicated to HFT trading in the last couple of years, leading to an "arms race" in progressively faster hardware and software, the profitability of high-frequency trading has declined. As the speed of transactions has increased, the ability to make money based on speed alone has diminished. Further increases in speed are now bringing steadily diminishing returns — the profit per transaction has plunged. As a result, successful trading now depends less and less on hardware and more on software in the form of sophisticated trading algorithms.

An *algorithm* is a set of instructions used to carry out a procedure, kind of like a recipe. Algorithms are heavily used by computer scientists to instruct computers on how to perform various tasks, such as carrying out mathematical operations.

The use of advanced algorithms for trading strategies carries several potential advantages, such as the ability to test ideas on historical data before risking any money. With HFT trading, there's no time to test any potential trading strategies, because they must be implemented immediately.

Another advantage to using trading algorithms is that they can be based on fundamental variables, such as interest rates and exchange rates, instead of simply searching through trades to look for temporary price changes. As a result, algorithms can be developed to find ever more complex relationships among securities prices and use this information to earn trading profits. Big data enhances algorithmic trading by providing the ability to search through enormous volumes of data looking for patterns that might not be detectable with smaller amounts of data or slower processing speeds.

With shrinking profits from HFT, algorithmic trading appears to have a bright future, as the increasing availability of data and computer speed enable more and more sophisticated algorithms to be developed.

Big Data and Electric Utilities

One area where big data has made an impact on electric utilities is the development of smart meters. *Smart meters* provide a more accurate measure of energy usage by giving far more frequent readings than traditional meters. A smart meter may give several readings a day, not just once a month or once a quarter.

The information gathered by these meters help customers conserve electricity (by providing them with a more accurate picture of their consumption patterns). It can also enable them to better plan their electricity usage to avoid peak hours and save money.

Smart meters also provide utilities with several advantages:

- More accurate forecasts of future energy demand
- Improvement in the scheduling of maintenance
- Increase in ability to detect fraud
- Reduction in power outages
- Better compliance with regulatory requirements

With smart meters, utilities can determine not only how much electricity is being used, but at what times of day it's being used. This information is critical in forecasting the demand for electricity at peak hours. Because electricity can't be stored, capacity must match use at peak hours — the rest of the time, much of this capacity remains idle. As a result, the more accurately utilities can measure peak demand, the more efficiently they can plan for capacity needs.

The biggest challenge to the utilities that use smart meters is that the amount of data being generated is dramatically greater than the amount generated by traditional meters. This fact requires a massive upgrade in the hardware and software capabilities of many utilities. Another problem is that the data being gathered may come from many different sources, leading to potential compatibility problems.

In the long run, the investments being made by utilities in big data capabilities may end up saving money by using existing resources more efficiently, thereby reducing the need to build new capacity.

Big Data and Higher Education

Big data is making dramatic changes in the field of education. One area that has shown particular promise is computerized learning programs, which provide instant feedback to educators. The data gathered from these programs can provide key information to identify key challenges:

- ✔ Students who need extra help
- ✔ Students who are ready for more advanced material
- ✔ Topics that students are finding especially difficult
- ✔ Different learning styles

This information enables educators to identify problem areas and come up with alternative methods for presenting material. Computerized testing can also be used to better understand the different techniques used by students to learn the same material. For example, some students do better with visual examples, but others do better with more numerically oriented examples. Educators can use data to customize training to the needs of individual students.

Additional advantages of using big data in education include the following:

- ✔ Improved ability to develop curricula that address student needs
- ✔ Development of customized instruction for each student
- ✔ Improvement of tools used to assess student performance

Several issues can arise with the use of big data in education. To use the data that's gathered, large investments may be required to upgrade hardware and software capabilities. Another potential issue is privacy concerns as student data becomes available to educators — clearly, safeguards need to be put into place to ensure that this data remains confidential.

Big Data and Retailers

Retailers collect and maintain sales records for large number of customers. The challenge has always been to put this data to good use. Ideally, a retailer would like to understand the demographic characteristics of its customers and what types of goods and services they are interested in buying. The continued improvement in computing capacity has made it possible to sift through huge volumes of data in order to find patterns that can be used to forecast demand for different products, based on customer characteristics.

Another issue that big data can help with is pricing strategies, specifically understanding how sensitive different customers are to prices. Choosing the right price for a product has sometimes been based on guesswork. By contrast, big data can increase the retailer's ability to use customer habits to identify the profit-maximizing price for their goods. Another benefit to using big data is that retail stores can better plan the placement of merchandise throughout the store, based on customer shopping habits.

Big data can also help retailers with inventory management. Many retailers sell a wide variety of different products, and keeping track of this information is a huge challenge. With big data, retailers can have instantly updated information about the size and location of their inventories.

One of the most important uses of big data for a retailer is the ability to target individual consumers with promotions based on his or her preferences. Such targeting not only increases the efficiency of advertising, it gives customers a more personal relationship with the retailer, thereby encouraging repeat business. In addition, knowledge of customer preferences enables the retailer to provide recommendations for future purchases, which further increases repeat business.

Nordstrom

As an example, Nordstrom has heavily embraced the use of big data. It was one of the first retail stores to offer customers the option of shopping online. The company has developed a smartphone app that lets customers shop directly from their iPads, iPhones, and other mobile devices. Nordstrom also shows customers which of its stores carries specific merchandise; for merchandise that must be ordered from other stores, Nordstrom can provide a highly accurate estimate of delivery time.

Nordstrom uses its big data capabilities to target customers with personalized ads based on their shopping experiences. This information can come from Nordstrom's store sales, its website, and from social media sites such as Facebook and Twitter.

Nordstrom conducts research into improving the customer shopping experience through its Innovation Labs division. It created this division in 2011 in order to ensure that the company remains on the cutting edge of big data technology.

Walmart

Walmart is another major retailer that has embraced big data. Based on sales volume, Walmart is the largest retailer in the United States. It's also the largest private employer in the country.

In the past few years, Walmart has made a major push into e-commerce, enabling it to compete directly with Amazon.com and other online retailers. In 2011, Walmart acquired a company called Kosmix to take advantage of that company's proprietary search engine capabilities (Kosmix was renamed Walmart Labs). Since then, Walmart Labs has developed several new products based on big data technology. One of these is called Social Genome, which enables Walmart to target individual customers with discounts based on preferences the customers have expressed through various sites on the Internet. Another product developed by Walmart Labs is Shoppycat, an app that provides gift recommendations based on information found on Facebook.

Although e-commerce still accounts for a relatively small percentage of Walmart's annual revenues, the investments the company has made in big data technology show that it expects online sales to become a progressively more important source of revenues in the future.

Amazon.com

The best example of using big data in retail is Amazon.com, which couldn't even exist without big data technology. Amazon started out selling books and has expanded into just about every area of retail imaginable, including furniture, appliances, clothes, and electronics. As a result, Amazon raked in $89 billion in revenues in 2014, making it one of the top ten retailers in the U.S., and the largest online retailer.

Like online retailers, Amazon uses big data for several applications:

- Managing its massive inventories
- Accurately keeping track of orders
- Making recommendations for future purchases

Amazon provides its recommendations through a process known as *item-to-item collaborative filtering.* This filtering is based on algorithms designed to identify the key details that can lead a customer to purchase a product, such as past purchases, items viewed, purchases made by customers with similar characteristics, and so forth. Amazon also provides recommendations by email, chosen based on the highest potential sales.

Amazon has been able to put its investment in big data capabilities to good use in another way: It now earns revenues by allowing businesses to use its infrastructure for a fee. This is done through products such as Amazon Elastic MapReduce (EMR) and Amazon Web Services (AWS).

Amazon EMR enables businesses to analyze enormous quantities of data by using Amazon's computer hardware. This hardware is accessible through the Amazon Cloud Drive, where businesses can pay to store their data. For many businesses, using these facilities is cheaper than building the computer infrastructure that would be required to handle the demands of big data. AWS provides a large variety of computer services through Amazon Cloud Drive, including storage facilities, database management systems, networking, and so on.

One interesting extension of Amazon's use of big data is its plan to ship merchandise to customers *before they order it!* The company received a patent in 2014 for its "anticipatory shipping" methodology. In order for this plan to succeed, Amazon.com must be able to anticipate customer demand with an incredibly high degree of accuracy to avoid the risk of returned merchandise.

Big Data and Search Engines

Big data has made possible the development of highly capable online search engines. A search engine finding web pages based on search terms requires sophisticated algorithms and the ability to process a staggering number of requests. Here are four of the most widely used search engines:

- Google
- Microsoft Bing
- Yahoo!
- Ask

The use of Google dwarfs its competitors. As of February 2015, Google is estimated to receive 1.1 billion unique visitors each month. Bing is a distant second with 350 million; Yahoo! gets 300 million, and Ask, 245 million. Although Google isn't the oldest search engine, it has become by far the most popular. The amount of data that Google handles each day is estimated to be about 20 petabytes (2.0×10^{16} bytes). All this traffic is profitable for Google — the bulk of its revenues come from advertising.

Google also provides computer services to organizations that don't have their own capabilities for processing big data. Google recently introduced Google Cloud Dataflow, which allows organizations to store, analyze, and process huge quantities of data.

Big Data and Social Media

Social media wouldn't be possible without big data. Social media websites let people share photos, videos, personal data, commentary, and so forth. Some of the best examples of social media websites include these:

- ✔ Facebook
- ✔ Twitter
- ✔ LinkedIn
- ✔ Instagram

Facebook was created in 2004 by Harvard students. It has since grown into the largest social media site in the world. As of 2014, Facebook has an estimated 900 million unique visitors each month, followed by Twitter with 310 million and LinkedIn with 225 million. Instagram has an estimated 100 million unique visitors per month.

Facebook allows users to stay informed about their friends, family members, acquaintances, and so on. Users can post photos, personal information, and news items, and use Facebook to locate friends. Twitter enables short messages called *tweets* to be posted. LinkedIn is a networking site intended to enable working professionals to stay in contact with each other, whereas the Instagram website lets users post photos for other users to see.

It's estimated that Facebook processes more than 500 terabytes of data every day, requiring an extensive investment in big data technology. Like Google, Facebook earns the bulk of its revenues from advertising, much of it from the largest corporations in the United States.

Regardless of what industry you're in, the techniques discussed in the remainder of this book will be helpful in your attempts to make smarter decisions. Hopefully, the examples given here have whetted your appetite for big data applications. And we hope you'll be able to sustain that appetite through the technical details that you'll come across. The chapters that follow start getting into the nuts and bolts of analyzing data.

Chapter 4

Understanding Probabilities

his chapter introduces some of the most important concepts related to the notion of probability. Probabilities underlie everything in the field of statistics. Having a clear sense of the basic ideas in probability theory will help you more easily digest the more advanced statistical ideas presented in this book.

The Core Structure: Probability Spaces

Beneath every statistical application, there lies a group of possible outcomes. A coin may come up heads or tails. A stock may go up or down. A baby may be born with blue, green, or brown eyes. These are generally referred to as *experiments*. In any situation where probability and statistics are applied, it is important to understand the entire set of possible outcomes. This set of possible outcomes is known as the *event space*. It is generally denoted by a capital E.

An event is a subset of the event space. An event can consist of more than one particular outcome. For example, the event that a stock advances by more than $10 on a given day consists of many specific outcomes. It could advance by $10.01, or $10.02, and so on. If any one of these outcomes occurs, we say the event occurs.

The second important thing to understand is the likelihood of various events occurring. For a fair coin flip, both heads and tails are equally likely and each is assigned a probability of 1/2. This assignment of probabilities to events is known as a *probability measure*. A probability measure is frequently represented by the letter P (not surprisingly). So, in the case of the coin flip example, we would write $P(H) = 1/2$ to represent the fact that the probability of obtaining heads is 50%.

We're being a little bit loose with the notion of an event space here. Technically, the set of all possible outcomes is called the *sample space*. The *event space* is actually derived from the sample space by assigning probabilities to various subsets of the sample space. Believe it or not, it is in general not possible to assign every subset of your event space a probability. So, the event space consists of a collection of *measurable* subsets of the sample space. However, those subsets that are not measurable are weird-looking beasts. We don't encounter them in the applications discussed in this book. From here on out, we'll treat the event space as the fundamental structure.

There are four basic properties that all probability measures share. They are all simple and quite intuitive:

- ✔ All probabilities are between 0 and 1.

- ✔ The probability of the entire event space is 1 or $P(E) = 1$.

- ✔ The probability of nothing happening is 0 or $P(\varnothing) = 0$. The symbol $\varnothing$ is generally used to represent the *empty set*.

- ✔ If two events don't overlap, then the probability of one or the other happening is the sum of their individual probabilities. Symbolically, if $A \cap B = \varnothing$, then it is always the case that $P(A \, or \, B) = P(A) + P(B)$. (The $\cap$ means intersection.)

From these simple rules, we can derive a great many useful results. Interestingly, and somewhat surprisingly perhaps, it is the ability to add probabilities for non-overlapping sets that prevents us from being able to assign probabilities to all subsets of the event space. It is possible, in some cases, to create a series of non-overlapping subsets that have a problematic property. One the one hand, they should add up to a probability of 1 because together they comprise the whole event space. But on the other hand, each one individually has probability 0. Strange but true.

Much of probability (and statistics) involves understanding the probability measure as just described, generally by looking at the probabilities associated with all the individual outcomes in the event space. This is known as the *probability distribution*.

There are two fundamentally different situations that you will encounter, depending on the structure of the event space. The next two sections discuss them and provide a couple of standard examples.

Discrete Probability Distributions

Many probability distributions have the property that each individual outcome in the event space has a positive (non-zero) probability. In this situation, the distribution is said to be *discrete*. It is possible to list all the outcomes with their associated probabilities.

This characterization may strike you as odd at first. Don't all outcomes have probabilities? The answer is no. There are many cases where this is not the case. There are distributions for which not a single individual outcome has a positive probability. In fact, the majority of statistical applications described in this book do not deal with discrete distributions. We explore that situation later in this chapter after introducing you to the discrete case.

A standard set of examples in probability involves flipping a coin one or more times. In the case of a single coin flip, the event space consists of only two individual outcomes. If the coin is fair, both heads and tails have a probability of 1/2. This completely describes the probability distribution of a single coin flip.

We said that with a discrete probability distribution it is possible to "list" the outcomes and their probabilities. But we're using the term loosely. Many discrete probability distributions have infinite event spaces. In these cases, *list* means to establish an easily understandable pattern that the probabilities follow.

One simple example of an infinite yet discrete probability is based on the experiment of flipping a coin until heads appears. In this case, the probability of heads on the first flip is 1/2. The probability of heads for the first time on the second flip is 1/4. On the third flip, it's 1/8 and so on. The probabilities get increasingly small. But they never quite go all the way to 0. As astronomically unlikely as it is, it is theoretically possible to toss tails a billion times in a row. (This sort of possibility was a theme in the cult classic movie *Rosencrantz and Gildenstern Are Dead*.)

This list of probabilities associated with each outcome is known as a *probability mass function*. There is another, related function known as the *cumulative mass function,* or *cumulative distribution function*. This function simply adds up the probabilities as you proceed across the event space from lowest to highest.

In the case of our first heads experiment, the value of the cumulative mass function at 1 is 1/2. But for 2, the cumulative probabilities are $1/2 + 1/4 = 3/4$. At 3, we get $1/2 + 1/4 + 1/8 = 7/8$. And so on.

The sections that follow take a look at some basic computational tools that are very helpful in calculating probabilities. Then we apply these tools to a commonly used classic discrete distribution called the *binomial distribution*.

Counting outcomes

Not all discrete distributions are as simple as the first heads experiment described in the preceding section. The probability of coming up with exactly five heads in ten coin flips is not quite so transparent. Luckily, there are a couple of mathematical techniques that are useful for answering questions like this.

Factorial: Counting how many ways you can arrange things

The exclamation point (!) is used to represent the mathematical operator *factorial*. You pronounce n! as "n factorial." So, what is n factorial? It's the product of all positive integers less than or equal to n. For example, consider the following:

$$0! = 1$$
$$1! = 1$$
$$2! = (2)(1) = 2$$
$$3! = (3)(2)(1) = 6$$
$$4! = (4)(3)(2)(1) = 24$$

 It might seem a little strange that 0! turns out to be 1. This is just a convention that mathematicians adopted to make the formulas work smoothly. It keeps them from having to put caveats in place when they write proofs. And it simplifies a lot of formulas.

You use the factorial operator to determine the number of ways that you can arrange a group of objects. For example, suppose you have a three-car garage, in which you are currently storing a Mercedes (M), Honda (H), and Infinity (I). In how many ways can you park these cars? The possibilities are as follows:

MHI   MIH   HMI   HIM   IMH   IHM

Each of these is an *arrangement* of the three cars; the list shows that there are six possible arrangements of three cars. You have three choices for the first car. This leaves two choices for the second. And then you only have one option for the third. This can be computed as $3! = 6$.

Combinations: Counting how many choices you have

In the preceding example regarding arrangements, the order in which the objects are placed matters. But suppose that, instead of arranging your cars in the garage, you've decided you want to downsize to one car. That means you need to pick two cars to sell. In this case, the order in which you pick them doesn't matter at all.

You use the formula

$$\frac{n!}{x!(n-x)!}$$

to determine the number of *combinations* that can be created when choosing x objects from a set of n objects. With a combination, the order in which objects are chosen is irrelevant.

One way to think of this formula is to think in terms of arrangements. You have $n!$ ways of arranging all the objects of interest. But order doesn't matter. So you could rearrange the x objects you chose in $x!$ different ways, and it wouldn't change the selection. In other words, that selection is repeated $x!$ times in the $n!$ arrangements. The same goes for the $(n-x)$ objects you didn't choose. By dividing the total number of arrangements by $x!$ and $(n-x)!$ you essentially remove the duplicate arrangements. This leaves you with the number of ways, or combinations, of choosing x objects from a set of n objects.

For example, you could use this formula to determine the number of ways that you can choose two electives from a set of six during the upcoming semester. The order in which the electives are chosen doesn't matter. Each possible selection is considered to be a *combination* of two electives chosen from six. You compute the number of choices as follows:

$$\frac{n!}{x!(n-x)!} = \frac{6!}{2!(6-2)!} = \frac{6!}{2!4!} = \frac{720}{(2)(24)} = 15$$

When only two things can happen: The binomial distribution

The binomial distribution is used in the following situations:

✓ A random process is taking place, consisting of a series of "trials."

✓ On each trial, one of two events could happen. These events are labeled "success" and "failure."

✓ The probability of a success on each trial is a fixed constant.

Under these conditions, you can use the binomial distribution to compute the probability that a specified number of successes occur.

Earlier we mentioned the difficulty of calculating the probability of getting exactly five heads in ten coin flips. This is precisely the sort of question that the binomial distribution is designed to answer.

It's sometimes a little confusing, to us at least, to think of "the" binomial distribution, because the probabilities depend on both the number of successes and the number of trials. We prefer to think of it as a set of distributions, one for each *n*. So we think of the binomial distribution related to 10 trials as distinct from the binomial distribution related to 100 trials. It's a minor point, but it helps keep our heads from spinning. A similar point could be made for the dependence on *p*, the probability of success.

Computing binomial probabilities

Let's compute binomial probabilities using a binomial formula. For example, you can determine the probability that five heads will turn up when a coin is flipped ten times. The binomial probability formula is this:

$$P(X = x) = \frac{n!}{x!(n-x)!} p^x (1-p)^{n-x}$$

In this formula:

X = A binomial random variable whose value is determined by the number of successes that occur during a series of trials

x = Number of successes that occur during a series of trials

n = Number of trials that take place

p = The probability of *success* on a single trial

$(1 - p)$ = The probability of *failure* on a single trial

! = The factorial operator

Don't confuse the variables X and x in this formula. The capital X is a binomial random variable, whereas the lowercase x is a specific value that refers to the number of successes whose probability you're calculating.

Binomial formula: Computing the probabilities

Combinations are useful for computing binomial probabilities. You can find the probability of x successes occurring during n trials with the binomial formula:

$$P(X = x) = \frac{n!}{x!(n-x)!} p^x (1-p)^{n-x}$$

For example, say 80 percent of all new cars have automatic transmissions, and the rest have manual transmissions. If a sample of five new cars is randomly chosen, what is the probability that four of them have automatic transmission?

The first step is to define automatic transmission as a *success*. Then in the binomial formula, $n = 5$ since five cars are chosen (that is, there are five trials of this experiment). $p = 0.80$ since this is the probability of automatic transmission (which is defined as "success"), and $x = 4$ since we are looking to compute the probability of four successes. Substituting these values into the binomial formula gives the following probability:

$$P(X = 4) = \frac{5!}{4!(5-4)!}(0.80)^4(0.20)^1 = 5(0.4096)(0.20) = 0.4096$$

In order to implement the binomial distribution in Excel, you can use the BINOM.DIST function for Excel 2010 or later; for Excel 2007 or earlier, the function is called BINOMDIST.

The values that you need to provide to the BINOM.DIST function are labeled number_s (the number of successes), trials, probability_s (the probability of a success on a single trial), and cumulative. This variable takes on a value of 1 for a "less than or equal to" probability, also known as a cumulative probability. For computing a single probability, you set cumulative equal to zero. Based on the example, $P(X = 4)$ is computed as BINOM.DIST(4, 5, 0.8, 0).

Moments of the binomial distribution

Moments are summary measures of a probability distribution. The *expected value* represents the mean or average value of a distribution. You obtain this by taking each possible value of the distribution, weighting it by its probability, and then summing the results. The *variance* and *standard deviation* represent the "dispersion" among the different possible values of a probability distribution. The formulas for computing the moments of the binomial distribution are given in the next two subsections.

Binomial distribution: Calculating the mean

The *expected value* of a probability distribution is its average value. You get it by weighting each possible value by its probability of occurring. For the binomial distribution, the calculation of the expected value can be simplified to the following:

$$E(X) = np$$

For example, suppose that 70 percent of all adults over 21 have at least a college education. Define "college graduate or better" as success. In a randomly chosen sample of 12 people, what is the expected number of college graduates? This is computed as follows:

$$E(X) = np = (12)(0.70) = 8.4$$

Binomial distribution: Computing variance and standard deviation

The *variance* of a distribution is the average squared distance between each possible outcome and the expected value. For the binomial distribution, you compute the variance with the following simplified formula:

$$\sigma^2 = np(1-p)$$

The *standard deviation* of a distribution equals the square root of the variance. For the binomial distribution, you compute the standard deviation as follows:

$$\sigma = \sqrt{np(1-p)}$$

For the education example,

- ✔ The variance is: $\sigma^2 = np(1-p) = 12(0.70)(0.30) = 2.52$.
- ✔ The standard deviation is the square root of 2.52, which is approximately 1.587.

Graphing the binomial distribution

You can illustrate discrete distributions, such as the binomial distribution, with a histogram. A *histogram* is a graph that consists of a series of vertical bars. Each bar represents a single value or a range of values for a random variable. The heights of the bars may be interpreted as either frequencies or probabilities.

For example, the binomial distribution for the education example is uniquely characterized by the number of trials $(n = 12)$ and the probability of success $(p = 0.7)$. Figure 4-1 shows a histogram that illustrates this distribution.

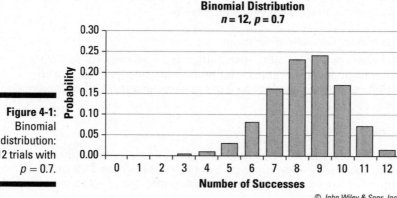

Figure 4-1: Binomial distribution: 12 trials with $p = 0.7$.

© John Wiley & Sons, Inc.

Continuous Probability Distributions

A *continuous probability distribution* is one for which you cannot assign probabilities to individual outcomes. The possible outcomes are distributed over some section of the real line. In the case of the normal distribution discussed in this section, they are distributed across the entire real line.

In this case, it is only possible to associate (nonzero) probabilities with intervals. In the case of continuous distributions, instead of probability mass functions for each outcome, we use a *probability density function* that reflects the varying probabilities of different intervals. Essentially, the probability that an outcome occurs in some interval is the area under the density function in that interval.

Perhaps the simplest example is the uniform distribution. This is what basic random number generators use. They generate a random number between 0 and 1 with all outcomes equally likely. In this case, the density function is constant on the interval from 0 to 1 and takes the value 1.

In this case, it is extremely easy to calculate probabilities associated with intervals. Probabilities are simply the length of the interval — the probability that a given random number falls between 0 and 1/2 is 1/2.

With continuous distributions, we also talk about cumulative distribution functions, which are analogous to the discrete case. They represent the probability that an outcome is less than a given value. This in fact is why they are useful in dealing with continuous distributions. Because all individual outcomes have probability 0, the values of the distribution function do not represent actual probabilities. The cumulative distribution function does, however, reflect an actual probability. In fact, you can use it to calculate the probability of any interval simply by subtracting the left endpoint from the right. We do a lot of that in later chapters.

Continuous probability distributions are useful for situations where measurements are involved. Heights, weights, return on investment, interest rates, and many other quantities of interest take values in some interval of the real line. Some of the most widely used continuous distributions in business applications are the normal, Student's t, lognormal, chi-square, and F-distributions. Most of these are addressed in detail in later chapters. The next section focuses on the normal distributions.

The normal distribution

The normal distribution is a continuous distribution that's widely used in business applications — and many other disciplines, such as psychology, sociology, biology, and so forth. The normal distribution is uniquely characterized by two values, known as *parameters:*

- ✔ The expected value (mean), represented by μ (the Greek letter *mu*)
- ✔ The standard deviation, represented by σ (the Greek letter *sigma*)

The normal distribution with a mean (μ) equal to 0 and standard deviation (σ) equal to 1 plays a special role in statistical analysis. It's often known as the *standard normal distribution.*

The expected value or mean of a probability distribution is known as the *first central moment* of the distribution. The variance is known as the *second central moment.* The standard deviation isn't a separate moment; it's simply the square root of the variance. Moments are useful for understanding the properties of a probability distribution.

One of the most important properties of the normal distribution is that it's *symmetrical* about its mean. This indicates that the area under the curve to the left of the mean is a *mirror image* of the area under the curve to the right of the mean. The symmetry of the distribution greatly simplifies the process of computing normal probabilities. You can see this in Figure 4-2, a diagram of the *standard normal distribution,* which is the special case in which the mean equals 0 and the standard deviation equals 1. The graph of this distribution is often referred to as a *bell-shaped curve* or just *bell curve.*

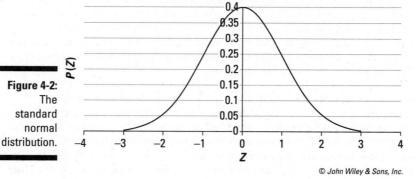

Standard Normal Distribution

Figure 4-2: The standard normal distribution.

© John Wiley & Sons, Inc.

By convention, the letter Z is used to represent the *standard normal* distribution, whereas the letter $X(\mu, \sigma)$ is used to represent any other normal distribution with mean μ and standard deviation σ.

Probabilities for the normal distribution correspond to areas under the bell-shaped curve. The entire area under the curve equals 1. Due to the symmetry of the bell-shaped curve, the area below the mean equals 0.5, as does the area above the mean.

Computing probabilities for the standard normal distribution

You can compute normal probabilities using a table, a specialized calculator, or one of Excel's built-in statistical functions. Normal tables are usually set up to show the probability that a normal random variable is *less than or equal to* a specified value. In other words, the table shows *cumulative* probabilities.

Table 4-1 shows a section of a table of *standard normal* probabilities.

Table 4-1	Standard Normal Table			
Z	0.00	0.01	0.02	0.03
0.9	0.8159	0.8186	0.8212	0.8238
1.0	0.8413	0.8438	0.8461	0.8485
1.1	0.8643	0.8665	0.8686	0.8708
1.2	0.8849	0.8869	0.8888	0.8907

As an example, the probability that a standard normal random variable (Z) is less than or equal to 1.12 is found at the intersection of 1.1 in the first column and the column headed 0.02; this equals 0.8686.

Other probabilities can be determined through the use of algebra and the properties of the standard normal distribution. For example, the probability that Z is *greater than* or equal to 1.12 equals $1 - 0.8686 = 0.1314$. This is because Z must be less than or equal to 1.12 or greater than or equal to 1.12. Therefore:

$$P(Z \leq 1.12) + P(Z \geq 1.12) = 1$$

$$P(Z \geq 1.12) = 1 - P(Z \leq 1.12) = 1 - 0.8686 - 0.1314$$

For computing probabilities with negative values, you can use a separate table. Table 4-2 shows a section of the standard normal table for negative values.

For example, the probability that Z is less than or equal to −1.02 is found at the intersection of −1.0 in the first column and the column headed 0.02; this equals 0.1539.

Table 4-2	Standard Normal Table — Negative Values			
Z	0.00	0.01	0.02	0.03
−1.3	0.0968	0.0951	0.0934	0.0918
−1.2	0.1151	0.1131	0.1112	0.1093
−1.1	0.1357	0.1335	0.1314	0.1292
−1.0	0.1587	0.1562	0.1539	0.1515

You can compute standard normal probabilities with the Excel function NORMSDIST. For example, to find the probability that Z is less than or equal to 1.50, the appropriate Excel command would be NORMSDIST(1.50), which equals 0.9332.

Computing non-standard normal probabilities

Many variables may follow the normal distribution but not follow the standard normal distribution. In these cases, you use a simple transformation to find the appropriate probability.

For example, suppose that X is a normal random variable with a mean of 5 and a standard deviation of 10. In order to compute the probability that X is less than or equal to 15, you may convert X to the standard normal form using the following equation:

$$Z = \frac{X - \mu}{\sigma}$$

The key terms in this expression are as follows:

Z = A standard normal random variable

X = A normal random variable with mean μ and standard deviation σ

In this case,

$$Z = \frac{X - \mu}{\sigma} = \frac{15 - 5}{10} = 1$$

Z is now a standard normal random variable. In this case, its value is 1 standard deviation above the mean. You can compute this probability from a standard normal table. Based on Table 4-1, the probability is 0.8413.

You can compute normal probabilities other than standard normal probabilities in Excel 2010 or later with the function NORM.DIST. In Excel 2007 or earlier, the function is NORMDIST.

NORMDIST requires four values:

- ✔ x (the value you are looking for)
- ✔ mean (the mean of the normal distribution you are using)
- ✔ standard_dev (the standard deviation of the normal distribution you are using)
- ✔ cumulative

The last value, cumulative, can only assume a value of 0 or 1. It is simply indicating whether you want to use the distribution function or the cumulative distribution function. Remember, the cumulative distribution function gives the probability of getting a value less than or equal to a given value. If you want a "less than or equal to" probability (that is, a cumulative probability), set this equal to 1. A value of 0 simply gives you the height of the normal curve at a given point. In the previous example, $P(X \le 15)$ would be computed as NORM.DIST(15, 5, 10, 1).

Introducing Multivariate Probability Distributions

Measures of association such as the covariance and correlation are based on a probability distribution that shows the probabilities of two different variables. This type of probability distribution is known as a *joint probability distribution.*

As an example of a joint probability distribution, Table 4-3 shows how many children are in each household in a small town, with X representing the number of boys in a household, and Y standing for the number of girls.

Table 4-3	Distribution of Children by Household				
	Y = 0	*Y = 1*	*Y = 2*	*Y = 3*	*Sum*
X = 0	0.2000	0.1500	0.1000	0.0125	0.4625
X = 1	0.1500	0.2000	0.0375	0.0000	0.3875
X = 2	0.1000	0.0375	0.0000	0.0000	0.1375
X = 3	0.0125	0.0000	0.0000	0.0000	0.0125
Sum	0.4625	0.3875	0.1375	0.0125	1.0000

Joint probabilities

The probability that two variables assume a specified value is known as a joint probability. In general, a *joint probability* can be expressed like this:

$$P(X = x, Y = y)$$

In this equation, the variable X equals a specified constant x, and Y equals a specified constant y. In Table 4-3, the probability of a specified combination of boys and girls is a joint probability. The joint probability is at the intersection of the appropriate row and column. For example, the probability that a household has one boy and one girl is expressed as follows:

$$P(X = 1, Y = 1) = 0.2000$$

This shows that the probability that a household has one boy $(X = 1)$ *and* one girl $(Y = 1)$ is 0.2000, or 20 percent.

Similarly, the probability that a household has two boys and no girls is expressed as follows:

$$P(X = 2, Y = 0) = 0.1000$$

This shows that the probability that a household has two boys $(X = 2)$ *and* no girls $(Y = 0)$ is 0.1000, or 10 percent.

Unconditional probabilities

An *unconditional probability* (also known as a *marginal probability*) is the probability that a variable equals a specified value. The following are examples of unconditional probabilities:

$$P(X = x)$$
$$P(Y = y)$$

Using the household example from the last section, you can compute the probability that a household contains one boy, regardless of the number of girls, as follows:

$$P(X = 1) = P(X = 1, Y = 0) + P(X = 1, Y = 1) + P(X = 1, Y = 2)$$
$$+P(X = 1, Y = 3)$$

This equals $0.1500 + 0.2000 + 0.0375 + 0.0000$, or 0.3875.

Similarly, you can compute the probability that a household contains no girls, regardless of the number of boys, as follows:

$$P(Y = 0) = P(X = 0, Y = 0) + P(X = 1, Y = 0) + P(X = 2, Y = 0)$$
$$+ P(X = 3, Y = 0)$$

This equals $0.2000 + 0.1500 + 0.1000 + 0.0125$, or 0.4625.

These probabilities can be easily determined from the joint probability table shown in Table 4-3. Notice the following:

- An unconditional probability for X equals the sum of a specified *row*.
- An unconditional probability for Y equals the sum of a specified *column*.

For example, the probability that a household contains one boy, regardless of the number of girls, equals 0.3875, which is the sum of the row labeled $P(X = 1)$. The probability that a household contains no girls, regardless of the number of boys, equals 0.4625, which is the sum of the column labeled $P(Y = 0)$.

Conditional probabilities

A *conditional probability* for a variable depends on knowledge of the value of a second variable. You express the conditional probability that X equals a specific value x given that Y equals a specific value y like this:

$$P(X = x \mid Y = y)$$

Going back to the joint probability table shown in Table 4-3, you are essentially pretending that the entire event space consists of a single column. To compute a conditional probability for X, use the following formula:

$$P(X = x \mid Y = y) = \frac{P(X = x, Y = y)}{P(Y = y)}$$

The numerator of this expression is the *joint* probability that $X = x$ and $Y = y$. The denominator is the *unconditional* probability that $Y = y$.

Similarly, to compute a conditional probability for Y, use the following formula:

$$P(Y = y \mid X = x) = \frac{P(X = x, Y = y)}{P(X = x)}$$

The numerator of this expression is the *joint* probability that $X = x$ and $Y = y$. The denominator is the *unconditional* probability that $X = x$.

Based on the household data in the previous example, you can calculate several conditional probabilities. For example, you compute the probability that a household has no boys *given that* it has two girls as follows:

$$P(X = 0 \mid Y = 2) = \frac{P(X = 0, Y = 2)}{P(Y = 2)} = \frac{0.1000}{0.1375} = 0.7273$$

This shows that among all households with two girls, in 72.73 percent of these households there are no boys.

As another example, you compute the probability that a household has one girl *given that* it has one boy as follows:

$$P(Y = 1 \mid X = 1) = \frac{P(X = 1, Y = 1)}{P(X = 1)} = \frac{0.2000}{0.3875} = 0.5161$$

This shows that among all households with one boy, in 51.61 percent of these households, there is one girl.

Whew! This has been an admittedly technical chapter, but it's quite an important one. The ideas discussed here are used repeatedly in the rest of the book. But we're not quite done yet with the foundational stuff. The next chapter looks at some of the key ideas related to statistical methods. Then we can start into the applications.

Chapter 5

Basic Statistical Ideas

· ·

In This Chapter

▶ Understanding the basic types of measurement

▶ Learning the fundamental statistical measures of central tendency

▶ Understanding hypothesis testing

· ·

T his chapter introduces some of the most important statistical concepts you need to get started with big data. It also introduces several summary measures that represent the key properties of a dataset.

A *dataset* may consist of the elements of a *population* of interest, or it may take the form of a sample. A *sample* is a *subset* of a population; it's chosen in such a way that it accurately represents the underlying population. For most empirical applications, sample data is used instead of population data due to the time and cost required to analyze an entire population.

When taking random samples, it is *critical* that the samples be really random. And that involves using a random number generator. It also involves making sure that each observation in your dataset is equally likely to be selected. Just taking the first 10 percent of the observations or selecting every tenth observation until your sample is as big as you want is almost always going to produce a sample that does not reflect the whole population.

Some Preliminaries Regarding Data

Not all data is created equal from the point of view of statistical procedures. For the purposes of this book, there are four types of data to be aware of. By *data types,* we don't mean the way the data is formatted or stored (though that is important and is discussed in later chapters). Here we are referring to the statistical properties of the data. These properties affect what sorts of statistical operations can be meaningfully applied to the data.

Nominal data

The simplest and least powerful, from a statistical point of view, is called *nominal* data. Nominal data does nothing more than categorize or differentiate observations. Examples include account numbers, product codes, airline flight numbers, and anything else that simply serves as an identifier.

Even though nominal data is numeric in many cases, the numbers are not measurements. There is no statistical or practical insight to be gained from investigation of the average flight number to arrive on a given day. The numbers themselves are arbitrary.

Ordinal data

The second type of data that you may encounter is what's called *ordinal* data. This is data that falls short of being a full-blown measurement in the sense of bank balances or interest rates. But it does put observations into a meaningful order. A simple example of this is the finishing order in the Kentucky Derby. After the race, each horse's name (a nominal variable) can be attached to its finishing position.

The critical shortcoming of ordinal data is that there is not a consistent difference between the various positions. The winner may beat second place by a nose. But the third place horse might be three lengths behind. This means that if you average the horses' placement over several races, you don't get a result that accurately compares the performance of the horses.

One of our pet peeves in this vein comes from websites and other survey-based customer satisfaction measures that are based on "a scale of 1 to 10" type evaluation. We're quite convinced that in almost all cases, there has been no attempt to verify that the differences between successive ratings reflect equal increments of "satisfaction." For this reason, an average customer satisfaction rating of 8.5 is not statistically meaningful. Yet these sorts of numbers get reported all the time.

Summary Statistical Measures

A *summary measure* is a numerical value that represents an important property of a sample or a population. For example, the average age of the residents of a city is a summary measure. The ages of all the residents are the population of interest. A randomly chosen selection of these ages is a *sample*. The average age of the sample of residents is an estimate of the average age of the entire population.

It is critical that the sample is selected randomly. The process that you use to sample needs to give every element of the population an equal chance of being selected. Otherwise, your sample may give misleading results. Also, the sample must be large enough to have a good chance of being representative of the population. A sample of one isn't likely to tell you much. More on sample size later in the book.

There are three types of summary measures. Each describes a different aspect of a dataset (whether a sample or population). Summary measures are classified as follows:

- ✔ Measures of central tendency
- ✔ Measures of dispersion
- ✔ Measures of association

Measures of central tendency

Measures of central tendency show the *middle* or *center* of a sample or population. For example, the average value of a dataset is a measure of central tendency. There are three widely used measures of central tendency, and each has advantages and disadvantages:

- ✔ Mode
- ✔ Median
- ✔ Mean

These three measures provide an instructive example of why understanding whether your data is nominal, ordinal, or interval is important. We address them in increasing order of complexity.

Mode

The *mode* of a dataset is simply the most commonly observed value. There is no formula for computing the mode, and the procedure is the same for both samples and populations. First sort the observations from the lowest to the highest. Then compute the *frequency* of each observation.

For example, suppose a sample consists of the following values:

0, 3, 8, 2, 2, 6, 5

The sorted values are as follows:

0, 2, 2, 3, 5, 6, 8

Because 2 appears twice in this sample, and no other value appears more than once, 2 is the mode.

One interesting feature of the mode is that unlike the mean or the median, there can be two or more modes, or no mode at all. For example, here's a sample with two modes:

1, 1, 3, 5, 7, 8, 8, 9

In this case, the modes are *both* 1 and 8, since these values appear twice in the sample, whereas none of the others is repeated more than once.

An example of a sample with no mode:

0, 2, 4, 6, 8, 10

None of the values is repeated, so there is *no* mode.

The mode can be useful for non-numeric data, when computing a mean or a median would be impossible. The mode is essentially the only statistical measure that can be correctly applied to nominal data.

Median

The *median* of a dataset is its midpoint. In other words, half of the observations are below the median, and half are above it. There is no one formula for computing the median; you compute it in two steps. The first step is to sort the observations from the lowest to the highest value. The next step is to identify the "central" observation in the dataset.

For example, suppose a sample consists of the following values:

1, 4, 2, 7, 5

Here are the sorted values:

1, 2, 4, 5, 7

The central value is the third value, or 4. Half of the observations are below 4 and half are above. Therefore, 4 is the median of this sample.

For an even number of observations, the procedure changes slightly. For example, suppose that a sample consists of the following values:

1, 4, 2, 7, 5, 6

Here are the sorted values:

1, 2, 4, 5, 6, 7

The third smallest and fourth smallest values are considered to be the two central values; in this case, they are 4 and 5. Take the midpoint between 4 and 5, which is 4.5. This is the median of the sample. Half of the observations in the sample (1, 2, 4) are below 4.5, and half of the observations (5, 6, 7) are above 4.5.

There is no difference in computing the median of a population. You compute both a sample and a population in exactly the same way.

The median can be a more meaningful measure than the mean when a dataset contains extremely large or small values, which are known as *outliers*. Outliers are the topic of Chapter 10.

Calculating the median requires you to be able to sort your data from lowest to highest. Obviously, then, your data needs to be ordered. As explained at the beginning of this chapter, ordering is exactly the property that characterizes ordinal data. So the median is a meaningful measure for ordinal data.

Mean

In statistics, the word *mean* is a synonym for *average*. You compute a mean by adding up all the elements in a dataset and then dividing by the number of elements. The equations in this section show how to compute the mean of a sample and the mean of a population.

Suppose a researcher chooses a sample of prices of a gallon of gas in a major city. The sample consists of the following eight prices:

$3.98, $4.19, $3.79, $3.99, $3.78, $3.69, $3.97, $4.13

He or she computes the mean price of this sample as follows:

$$\bar{X} = \frac{\sum_{i=1}^{n} X_i}{n} = \frac{3.98 + 4.19 + 3.79 + 3.99 + 3.78 + 3.69 + 3.97 + 4.13}{8} = \frac{31.52}{8} = 3.94$$

This equation shows that the sum of the prices is $31.52. When divided by the sample size of 8, this gives a mean of $3.94 per gallon.

If the eight gas stations that were randomly chosen reflect the underlying population of gas prices, then the sample mean of $3.94 per gallon will provide a good estimate of the mean price for the entire population.

In general, the mean of a sample is computed as follows:

$$\bar{X} = \frac{\sum_{i=1}^{n} X_i}{n}$$

This formula uses the following terms:

$\bar{X}$ (pronounced "X bar") is the mean of the sample.

Σ (the upper case Greek letter sigma) is used to indicate that a sum is being computed.

n is the number of elements in the sample.

i is an *index* used to assign a number to each sample element, ranging from 1 to n.

X_i is a single element in the sample.

The mean of a population is computed as follows:

$$\mu = \frac{\sum_{i=1}^{N} X_i}{N}$$

The new term in this formula is μ, which is the Greek letter *mu*. It represents the mean of a population. Also note here the convention of using a capital N to represent the population size. This helps to avoid confusion with the sample size.

It's common practice in statistics to use the Greek alphabet to represent summary measures of populations, and the English alphabet to represent summary measures of samples.

Calculation of the mean (and in fact all the other measures discussed in this chapter) requires that your data meet the requirements of interval data. Remember, interval data represents measurements that can be compared. Averages aren't meaningful if your observations can't be compared at face value.

See Chapter 4 for info on the normal distribution. This distribution has some very nice properties from a mathematical standpoint. One of the things that makes it "normal" is that the three measures described in this section coincide. In other words, if a distribution is normal, then the mean, median, and mode are all equal.

Measures of dispersion

Measures of dispersion show how *spread out* the elements of a sample or population are — in other words, how close the elements are to each other. For example, the elements of the sample

0, 1, 1, 2, 2, 3, 4, 5

are very close to each other, which means that the spread between them is very small. On the other hand, the elements of the sample

–200, –50, 0, 50, 100, 150, 200

are far apart from each other. Therefore, the spread between them is significantly larger.

In many applications, it's important to compare the spread for different samples or populations. For example, the spread between the returns to a portfolio of stocks is considered to be a measure of risk. In order to compare the risk of different stock portfolios, a measure of dispersion is needed.

Here are three of the most important measures of dispersion:

✔ Range

✔ Variance

✔ Standard deviation

Range

The *range* of a dataset is simply the difference between its largest and smallest elements. This holds for both samples and populations. For example, if a sample consists of the elements

1, 2, 2, 7, 9, 10

then the range is $10 - 1$, or 9.

A drawback to this measure is that range only considers the values of the smallest and largest elements, without considering the spread between the individual elements. That makes range hypersensitive to outliers. For many applications, the variance and standard deviation are more useful measures of dispersion.

Variance

The *variance* equals the average squared difference between the elements of a dataset and the mean value of the dataset. The more spread out the elements of the dataset, the larger the variance is.

For example, suppose a sample of the number of residents per household in a small city is taken, and the results are as follows:

2, 1, 3, 4, 2, 6

The first step is to compute the sample mean:

$$\bar{X} = \frac{\sum_{i=1}^{n} X_i}{n} = \frac{2+1+3+4+2+6}{6} = \frac{18}{6} = 3$$

Then substitute the result (3) into the sample variance formula as follows:

$$s^2 = \frac{(2-3)^2 + (1-3)^2 + (3-3)^2 + (4-3)^2 + (2-3)^2 + (6-3)^2}{5} = \frac{16}{5} = 3.2$$

For each element, you compute the squared difference between the element and the sample mean. You divide the sum of these squared differences by one less than the sample size $(6-1=5)$ to get a sample variance of 3.2.

The following equations show how to compute the variance of a sample and a population.

You compute the variance of a sample like this:

$$s^2 = \frac{\sum_{i=1}^{n} (X_i - \bar{X})^2}{n-1}$$

The key terms in this formula are as follows:

s^2 = The sample variance

X_i = A single element in the sample

$\bar{X}$ = The sample mean ("X bar")

n = The number of elements in the sample

You compute the variance of a population as follows:

$$\sigma^2 = \frac{\sum_{i=1}^{N}(X_i - \mu)^2}{N}$$

The key terms in this formula go like this:

σ^2 = The population variance

X_i = A single element in the sample

μ = The population mean

N = The number of elements in the population

You might notice that unlike the sample and population mean formulas, there is a small algebraic difference between the calculations of the sample variance and the population variance. This is due to the fact that once the sample mean is known, you can calculate the value of the nth observation based on the sample mean and the other $n-1$ observation. The mathematical consequence of this is that if you divide by n instead of $n-1$, you will under-estimate the population variance.

σ is the lowercase Greek letter *sigma*. The uppercase sigma is Σ.

One of the drawbacks to the variance measure is that it's measured in *squared units*. In this example, the variance is actually measured in units of *squared residents!* Since this is extremely awkward to analyze, an alternative measure known as the *standard deviation* is more commonly used to measure dispersion.

Standard deviation

The *standard deviation* is simply the square root of the variance. This ensures that the deviation of a dataset is measured in the same units as the dataset, instead of squared units. Using the example of the number of residents per household, here's the sample standard deviation:

$$\sqrt{3.2} = 1.79$$

The standard deviation is 1.79, which indicates how much the number of residents per household can vary from one household to the next.

Overview of Hypothesis Testing

Many tests of the assumptions about a dataset take the form of *hypothesis tests*. Hypothesis testing is a six-step procedure designed to determine if an assumed statement is true or false.

Here are the six steps required to test a hypothesis:

- ✔ Null hypothesis
- ✔ Alternative hypothesis
- ✔ Level of significance
- ✔ Test statistic
- ✔ Critical value(s)
- ✔ Decision

This section discusses all of these in turn.

The null hypothesis

The *null hypothesis* is a statement that's assumed to be true unless there is *strong* evidence against it. In that case, the null hypothesis is *rejected*.

You can test many different types of null hypotheses. For example, the null hypothesis may be a statement about any of the following:

- ✔ The mean value of a population
- ✔ The variance of a population
- ✔ The mean difference between two different populations
- ✔ The probability distribution followed by a population

For example, if the null hypothesis is that the population mean equals a specified value, it would be written as follows:

$$H_0: \mu = \mu_0$$

The key terms in this expression are:

H_0 = The null hypothesis

μ = The population mean

μ_0 = The *hypothesized* value of the population mean

The alternative hypothesis

The *alternative hypothesis* is a statement that is accepted if the null hypothesis is rejected. Depending on the type of null hypothesis being tested, the alternative hypothesis can typically be implemented in one of three forms:

- ✔ Right-tailed test
- ✔ Left-tailed test
- ✔ Two-tailed test

For example, say the null hypothesis is that the population mean equals a specific value. Table 5-1 shows the corresponding alternative hypotheses that may be used.

Table 5-1	Alternative Hypotheses
	Alternative Hypothesis
Right-tailed test	$H_1: \mu > \mu_0$
Left-tailed test	$H_1: \mu < \mu_0$
Two-tailed test	$H_1: \mu \neq \mu_0$

H_1 = The alternative hypothesis

The table shows that with a right-tailed test, the alternative hypothesis is that the mean is *greater than* the specified value. With a left-tailed test, the alternative hypothesis is that the mean is *less than* the specified value. With a two-tailed test, the alternative hypothesis is that the mean is *unequal to* (either less than or greater than) the specified value.

The level of significance

The *level of significance* of a hypothesis test refers to the probability of rejecting the null hypothesis when it is actually true. In statistical jargon, this is known as a *Type I error*. A *Type II error* occurs when you fail to reject the null hypothesis when it's actually false.

The smaller the level of significance, the less likely there is to be a Type I error, but the more likely there is to be a Type II error. The level of significance is selected based on the relative importance of avoiding a Type I error compared with a Type II error. In many applications in finance and economics, the level of significance is chosen to be 0.05 (5 percent).

The test statistic

A *test statistic* is a numerical measure that is computed from sample data to determine whether or not the null hypothesis should be rejected. The form of the test statistic depends on the type of null hypothesis that is being tested. For example, if a hypothesis about the mean of a population is being tested, and a *small* sample of data (fewer than 30 observations) is drawn from the population, then the appropriate test statistic is as follows:

$$t = \frac{\bar{X} - \mu_0}{s / \sqrt{n}}$$

In this formula,

t indicates that this test statistic follows the Student's t-distribution.

$\bar{X}$ is the sample mean.

μ_0 is the hypothesized value of the population mean.

s is the sample standard deviation.

n is the sample size.

If the sample size is *large* (30 or more observations), and the population standard deviation is unknown, the test statistic becomes as follows:

$$Z = \frac{\bar{X} - \mu_0}{s / \sqrt{n}}$$

Z indicates that this test statistic follows the standard normal distribution.

The critical value (s)

In order to determine whether a hypothesis should be rejected, the test statistic is compared with one or two *critical values*. The critical values depend on the type of hypothesis being tested as well as the alternative hypothesis being used.

For example, when testing a hypothesis about the mean about the population with a *small* sample, the critical values are determined as follows:

- ✔ **Two-tailed test:** Critical value = $\pm t_{\alpha/2}^{n-1}$
- ✔ **Right-tailed test:** Critical value = t_{α}^{n-1}
- ✔ **Left-tailed test:** Critical value = $-t_{\alpha}^{n-1}$

In these formulas, α is the level of significance; n represents the sample size. These critical values are drawn from the *Student's t-distribution* with $n-1$ *degrees of freedom (df)*. The number of degrees of freedom used with the t-distribution depends on the particular application. For testing hypotheses about the population mean, the appropriate number of degrees of freedom is one less than the sample size (that is, $n-1$).

The critical value or values represent *tail areas* under the student's t-distribution. For a two-tailed test, the value of the level of significance (α) is split in half; the area in the right tail equals $\alpha/2$ and the area in left tail equals $\alpha/2$, for a total of α.

When testing a hypothesis about the mean about the population with a *large* sample, the critical values are determined as follows:

- ✔ **Two-tailed test:** Critical value = $\pm Z_{\alpha/2}$
- ✔ **Right-tailed test:** Critical value = Z_{α}
- ✔ **Left-tailed test:** Critical value = $-Z_{\alpha}$

These critical values are drawn from the *standard normal distribution* which we introduce in Chapter 4.

To reject or not to reject, that is the question

You can make the decision whether or not to reject the null hypothesis in one of two ways:

- ✔ Compare the test statistic with the critical value(s).
- ✔ Compare the *probability value (p-value)* to the level of significance.

Comparing the test statistic with the critical value(s)

You can determine whether the null hypothesis should be rejected as follows:

- **Right-tailed test:** If the test statistic is *greater than* the critical value, reject the null hypothesis H_0: $\mu = \mu_0$ in favor of the alternative hypothesis H_1: $\mu > \mu_0$; otherwise, do *not* reject the null hypothesis.

- **Left-tailed test:** If the test statistic is *less than* the critical value, reject the null hypothesis H_0: $\mu = \mu_0$ in favor of the alternative hypothesis H_1: $\mu < \mu_0$; otherwise, you do *not* reject the null hypothesis.

- **Two-tailed test:** If the test statistic is *less than* the negative critical value, reject the null hypothesis H_0: $\mu = \mu_0$ in favor of the alternative hypothesis H_1: $\mu < \mu_0$. If the test statistic is *greater than* the positive critical value, reject the null hypothesis H_0: $\mu = \mu_0$ in favor of the alternative hypothesis H_1: $\mu > \mu_0$. Otherwise, you do *not* reject the null hypothesis.

Comparing the p-value to the level of significance

As an alternative to comparing the test statistic with critical values, you can compare the p-value with the level of significance. The p-value represents the likelihood that the test statistic has a specified value when the null hypothesis is true. A very small p-value indicates that the null hypothesis is unlikely to be true.

The decision rule when using the p-value is as follows:

If the p-value is *below* the level of significance, reject the null hypothesis; otherwise, do not reject the null hypothesis.

If you reject the null hypothesis when using a two-tailed test, the sign of the test statistic determines which alternative hypothesis to accept. If the test statistic is negative, the left-tailed alternative hypothesis is accepted; if the test statistic is positive, the right-tailed alternative hypothesis is accepted.

Measures of association

Measures of association refers to the relationship between the elements of *two* datasets. As an example, it may be useful to determine whether there's a strong association between speed limits and highway accidents in different states. This type of analysis requires using a numerical measure that describes how closely two variables follow each other.

Two of the most important measures of dispersion are covariance and correlation.

Covariance

Covariance measures the degree to which the elements in two datasets tend to follow each other. For example, suppose one dataset consists of the annual sales of a corporation, and the other dataset consists of the same corporation's annual profits. There is a strong tendency for profits to rise during years when sales are strong, and fall when sales are weak. As a result, the covariance between sales and profits tends to be very high. On the other hand, the covariance between profits and labor costs would tend to be very low or even negative, since rising labor costs reduce profits.

For example, suppose you choose a sample of annual returns to Stock X and Stock Y for the last six years. Table 5-2 shows these returns.

Table 5-2	Returns to Stocks X and Y for 2009–2014	
Year	*Stock X*	*Stock Y*
2009	0.05	−0.02
2010	0.03	0.01
2011	0.02	0.05
2012	0.09	0.04
2013	−0.02	0.07
2014	0.01	0.06

A investor wants to analyze how closely the returns to Stocks X and Y follow each other. In other words, the investor wants to know whether both stocks tend to do well during the same years and do poorly during the same years. The covariance can determine this.

If the covariance between the stocks is positive, then there's a tendency for the two stocks to do well at the same time and to do poorly at the same time. If the covariance is negative, then the two stocks tend to do well during different years and do poorly during different years. If the covariance is zero (or very close to it), then there is no connection between the returns to the two stocks.

The first step in computing the covariance is to calculate the sample mean for both Stock X and Stock Y:

$$\bar{X} = \frac{\sum_{i=1}^{n} X_i}{n} = \frac{0.05 + 0.03 + 0.02 + 0.09 - 0.02 + 0.01}{6} = \frac{0.18}{6} = 0.03$$

$$\bar{Y} = \frac{\sum_{i=1}^{n} Y_i}{n} = \frac{-0.02 + 0.01 + 0.05 + 0.04 + 0.07 + 0.06}{6} = \frac{0.21}{6} = 0.035$$

These equations show that the average return to Stock X is 0.03 (or 3 percent), and the average return to Stock Y is 0.035 (or 3.5 percent).

Substitute these results into the sample covariance formula. The numerator of the sample covariance formula is computed as follows:

$$\sum_{i=1}^{n}(X_i - \bar{X})(Y_i - \bar{Y}) = (0.05 - 0.03)(-0.02 - 0.035) + (0.03 - 0.03)(0.01 - 0.035)$$
$$+ (0.02 - 0.03)(0.05 - 0.035) + (0.09 - 0.03)(0.04 - 0.035)$$
$$+ (-0.02 - 0.03)(0.07 - 0.035) + (0.01 - 0.03)(0.06 - 0.035)$$
$$= -0.0032$$

Divide the result by one less than the common sample size for Stocks X and Y (6 years) to give the following result:

$$s_{XY} = \frac{\sum_{i=1}^{n}(X_i - \bar{X})(Y_i - \bar{Y})}{n-1} = \frac{-0.0032}{5} = -0.00064$$

Because the covariance is negative, this shows that Stocks X and Y tend to do well in different years and poorly in different years. For example, Stock X had a strong return of 0.05 (5 percent) in 2009, while Stock Y had a loss of 0.02 (2 percent) during the same year. In 2013, Stock Y had a return of 0.07 (7 percent) while Stock X had a loss of 0.02 (2 percent).

The following equations show how to compute the sample covariance in general.

You compute the covariance between two samples like this:

$$s_{XY} = \frac{\sum_{i=1}^{n}(X_i - \bar{X})(Y_i - \bar{Y})}{n-1}$$

The key terms in this formula are as follows:

s_{XY} is the sample covariance between variables X and Y (the two subscripts indicate that this is the sample covariance, not the sample standard deviation).

$\bar{X}$ is the sample mean for X ("X bar").

$\bar{Y}$ is the sample mean for Y ("Y bar").

n is the number of elements common to both samples.

i is an *index* that is used to assign a number to each sample element, ranging from 1 to n.

X_i is a single element in the sample for X.

Y_i is a single element in the sample for Y.

The covariance cannot be calculated unless the sample size is the same for both variables. n refers to this size.

Correlation

The *correlation* measure is closely related to covariance. Correlation is often preferred as a measure of association for two main reasons:

- Unlike covariance, correlation can only assume values between –1 and 1, so its value can be more easily interpreted.

- Covariance is affected by the units in which it's measured, whereas correlation has no units.

You can compute the correlation between Stocks X and Y in the preceding example by dividing the sample covariance by the product of the sample standard deviations of Stock X and Stock Y.

The sample variance of Stock X is as follows:

$$s_X^2 = \frac{\begin{array}{c}(0.05-0.03)^2 + (0.03-0.03)^2 + (0.02-0.03)^2 + (0.09-0.03)^2 \\ + (-0.02-0.03)^2 + (0.01-0.03)^2\end{array}}{5}$$

$$= \frac{0.0070}{5}$$

$$= 0.0014$$

This equation shows that the variance of the returns to Stock X is 0.0014. The standard deviation is the square root of the variance:

$$s_X = \sqrt{0.0014} = 0.03742$$

The standard deviation of the returns to Stock X is 0.03742, or 3.742 percent.

The sample variance and standard deviation of the returns to Stock Y are computed as follows:

$$s_Y^2 = \frac{\begin{array}{c}(-0.02-0.035)^2 + (0.01-0.035)^2 + (0.05-0.035)^2 + (0.04-0.035)^2 \\ + (0.07-0.035)^2 + (0.06-0.035)^2\end{array}}{5}$$

$$= \frac{0.00575}{5}$$

$$= 0.00115$$

This equation shows that the variance of the returns to Stock Y is 0.00115. The standard deviation is the square root of the variance:

$$s_Y = \sqrt{0.00115} = 0.03391$$

The standard deviation of the returns to Stock Y is 0.03391, or 3.391 percent.

Substitute these results into the sample correlation formula:

$$r_{XY} = \frac{s_{XY}}{s_X s_Y} = \frac{-0.00064}{(0.03742)(0.03391)} = -0.50439$$

The result shows a strong tendency for the returns to Stock X and Stock Y to move in opposite directions. In other words, during years in which Stock X's returns are strong, Stock Y's returns tend to be weak, and vice versa.

Depending on the situation, you will see correlation numbers across the spectrum from −1 to 1. Numbers very close to either −1 or 1 imply that there is a near linear relationship between X and Y. In other words, one is a multiple of the other. This can actually be a problem in situations where you are trying to build a predictive model (read more about this in Chapter 15).

Here is the general formula for computing the correlation of a sample:

$$r_{XY} = \frac{s_{XY}}{s_X s_Y}$$

The key terms in this formula are as follows:

r_{XY} is sample correlation between X and Y.

s_{XY} is sample covariance between X and Y.

s_X is sample standard deviation of X.

s_Y is sample standard deviation of Y.

Higher-Order Measures

In addition to the measures of central tendency, dispersion, and association, there are other statistical measures that can be extremely useful. Two of these are skewness and kurtosis.

These measures are useful in several disciplines, including portfolio management and risk management. For example, skewness can be used by portfolio managers to determine whether an asset or portfolio of assets is

- ✔ Equally likely to earn a positive or negative return.
- ✔ More likely to earn a positive return than a negative return.
- ✔ More likely to earn a negative return than a positive return.

Kurtosis can be used in risk management to determine the likelihood of extreme outcomes for a portfolio (for example, large losses or large gains) compared with an assumed distribution such as the normal.

Skewness

Skewness measures the degree of *asymmetry* of a probability distribution. With a *symmetrical* distribution, the values of a variable are evenly distributed above and below the mean. Figure 5-1 shows the returns to a sample portfolio over the past 20 years. These returns are symmetrical about the mean — losses are equally likely as gains of the same magnitude.

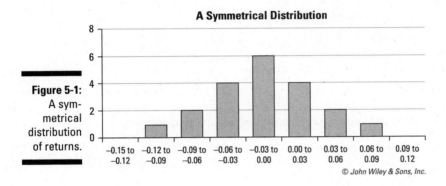

A Symmetrical Distribution

Figure 5-1:
A symmetrical distribution of returns.

© John Wiley & Sons, Inc.

Figure 5-2 shows the returns to another portfolio over the same time horizon. For this portfolio, there were more negative returns than positive returns over the past 20 years. As a result, the distribution is *negatively skewed* (or *skewed to the left*).

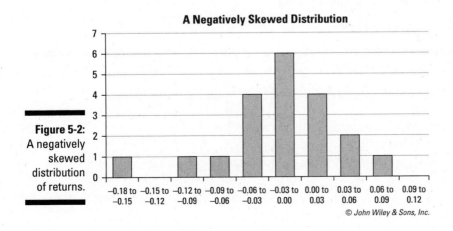

A Negatively Skewed Distribution

Figure 5-2:
A negatively skewed distribution of returns.

© John Wiley & Sons, Inc.

Figure 5-3 shows the returns to another portfolio over the same time horizon. For this portfolio, there were more positive returns than negative returns over the past 20 years. As a result, the distribution is *positively skewed* (or *skewed to the right*).

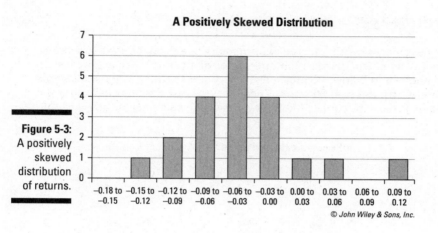

Figure 5-3: A positively skewed distribution of returns.

© John Wiley & Sons, Inc.

You compute the skewness measure for a sample as follows:

$$\frac{\frac{1}{n}\sum_{i=1}^{n}(X_i - \bar{X})^3}{\left[\frac{1}{n-1}\sum_{i=1}^{n}(X_i - \bar{X})^2\right]^{\frac{3}{2}}}$$

The key terms in this formula are as follows:

X_i = A single element in the sample

$\bar{X}$ = The sample mean ("X bar")

n = The sample size

Unlike mean, variance, and standard deviation, there is no universal symbol to represent skewness.

You compute the skewness measure for a population as follows:

$$\frac{\frac{1}{n}\sum_{i=1}^{n}(X_i - \mu)^3}{\left[\frac{1}{n}\sum_{i=1}^{n}(X_i - \mu)^2\right]^{\frac{3}{2}}}$$

The key terms in this formula are as follows:

X_i = A single element in the sample

μ = The population mean

n = The sample size

Kurtosis

Kurtosis measures the likelihood of extreme outcomes in a distribution relative to the normal distribution. The value of the kurtosis measure for the normal distribution is 3.

For a probability distribution with a kurtosis *above* 3, extreme outcomes are *more* likely to occur than with the normal distribution. This type of distribution is said to be a *fat-tailed* distribution, because more area is in the tails of the distribution and less area is near the mean than is the case with the normal distribution.

For a probability distribution with a kurtosis *below* 3, extreme outcomes are *less* likely to occur than with the normal distribution. This type of distribution is said to be a *skinny-tailed* distribution, because less area is in the tails of the distribution and more area is near the mean than with the normal distribution.

You compute kurtosis for a sample as follows:

$$\frac{\frac{1}{n}\sum_{i=1}^{n}(X_i-\bar{X})^4}{\left[\frac{1}{n-1}\sum_{i=1}^{n}(X_i-\bar{X})^2\right]^2}$$

The key terms in this formula are as follows:

X_i = A single element in the sample

$\bar{X}$ = The sample mean ("X bar")

n = The population size

As with skewness, there's no universally used symbol to represent kurtosis.

The kurtosis measure is computed for a population as follows:

$$\frac{\frac{1}{n}\sum_{i=1}^{n}(X_i - \mu)^4}{\left[\frac{1}{n}\sum_{i=1}^{n}(X_i - \mu)^2\right]^2}$$

The key terms in this formula are as follows:

X_i = A single element in the sample

μ = The population mean

n = The population size

Excess kurtosis is a measure that's sometimes used instead of kurtosis. *Excess kurtosis* is defined as kurtosis minus 3. Excess kurtosis has the advantage that positive excess kurtosis indicates a fat-tailed distribution, whereas negative excess kurtosis indicates a skinny-tailed distribution. The normal distribution has an excess kurtosis of 0.

Figure 5-4 shows a fat-tailed distribution of portfolio returns. The distribution has more extreme returns (positive and negative) than would be the case with the normal distribution.

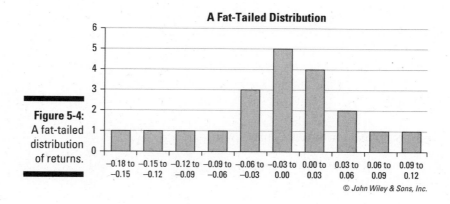

A Fat-Tailed Distribution

Figure 5-4:
A fat-tailed distribution of returns.

© John Wiley & Sons, Inc.

Figure 5-5 shows a skinny-tailed distribution of portfolio returns. The distribution has fewer extreme returns (positive and negative) than would be the case with the normal distribution.

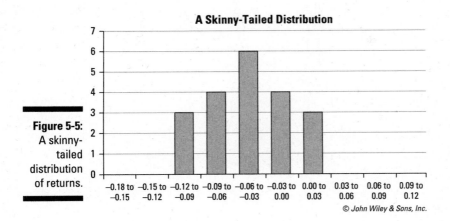

Figure 5-5:
A skinny-tailed distribution of returns.

And now we have finally come to the end of our whirlwind tour of probability and statistics concepts. If you made it through unscathed, you are to be congratulated.

Now it's time to get down to the business at hand, and learn how to use these concepts. In the next part, you will see how to apply these ideas to the process of data preparation. After that, you'll see how what you've learned so far can be applied in understanding your big data and predicting significant events.

Part II
Preparing and Cleaning Data

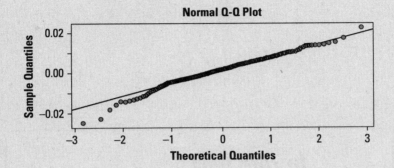

Check out an article on discrete and continuous probability distributions at
www.dummies.com/extras/statisticsforbigdata.

In this part . . .

- ✔ Getting your data ready for analysis
- ✔ Becoming familiar with popular computer file formats
- ✔ Understanding how to test for normality
- ✔ Checking out techniques for dealing with missing or incomplete data
- ✔ Knowing what to do about outliers

Chapter 6

Dirty Work: Preparing Your Data for Analysis

"*G*arbage in, garbage out" (GIGO) is a cliché that dates back to the early days of data processing. It succinctly captures the idea that any analysis that you do is only as good as the data you start with. In the context of statistical analysis, GIGO is particularly relevant. It is very easy to get caught up in the apparent power of statistical methods and software packages. They can seem almost magical in their predictive powers. But these predictions can be inaccurate and sometimes wildly misleading if the data they are analyzing is messy.

And the data is almost always messy — especially when it is *big data*. As discussed in Chapter 2, big data can be characterized by the three Vs: volume, velocity, and variety. All three of these characteristics open data up to problems.

The high volume and velocity of data open it up to technical problems with the way it is captured. Database systems don't always update properly. Websites crash and need to be reset. Systems of all sorts need to be taken down for maintenance.

Also, over time, the data may change in significant ways. A stock splits two for one, and suddenly the stock price is cut in half. A new product is introduced at a retailer, requiring a new code to be introduced. Or worse yet, a code may be recycled from a discontinued product.

The wide variety of data is also potentially problematic. Most large analytic databases are sourced from a variety of different systems. These systems record data in different ways. Date formats are not consistent from country to country. A customer's name spelled slightly differently in two different systems creates the potential for duplicate records.

All this messiness is collectively referred to as *dirty data*. This chapter introduces you to some common methods for dealing with dirty data. As the title suggests, cleaning up your data is not the most glamorous job. And it is often the most time-consuming part of doing statistical analysis. But it is absolutely critical to getting meaningful insight from your analytic efforts.

Throughout this chapter and as you apply the methods described in this book, keep in mind that there is not a "one-size-fits-all" answer to how to deal with dirty data. Having a clear focus on what you are trying to do is important. Your analytical goals will inform your decisions about what to do with regard to data problems.

Passing the Eye Test: Does Your Data Look Correct?

Most datasets come with some sort of *metadata,* which is essentially a description of the data in the file. Metadata typically includes descriptions of the formats, some indication of what values are in each data field, and what these values mean.

When you are faced with a new dataset, never take the metadata at face value. The very nature of big data requires that the systems that generate it are kept up and running as much as possible. For this reason, updating the metadata for these systems when changes are implemented is not always a top priority. You need to confirm that the data really is as the metadata claims.

Checking your sources

As obvious as it may sound, it is important that you have faith in where your data is coming from. This is particularly important when you are purchasing data. Thousands of vendors out there offer every imaginable kind of data. And they are not all of equal credibility.

Before purchasing data, try to understand exactly where and how the vendor is collecting it. Mysteriousness and vagueness are red flags.

Don't take vendors at their word. Don't rely solely on customer satisfaction postings on the website or client references provided by the vendor. If possible, try to track down someone who is using or has used the data.

If your data is coming from internal systems, it's still important to evaluate the sources. Different systems have different purposes and therefore focus on different data. They may also collect data at different times.

For example, it is not uncommon for some hotel chains to book reservations in a separate system from the one they use at the front desk when the guest checks in. It is possible that the guest may receive a discounted offer between booking and check-in. This means that the room rate in the reservation system may not match the rate in the front desk system. What's more, the reservation might get cancelled and never make it to the front desk!

Now, suppose you're performing an analysis of hotel revenues by city. It's rather important that you know that your room rate data is being sourced from the front desk system rather than the reservation system. But what if you're trying to analyze how many reservations were generated by your company's Super Bowl commercial? In this case, you want to see data from the reservation system.

The hotel example illustrates that even intrinsically clean data can be problematic. Even if data is accurate and exactly what it purports to be, timing can be an issue. Data changes over time.

Verifying formats

As mentioned earlier in this chapter, one of the things that your metadata will provide for you is some indication of how the data is formatted. By *formatted,* we mean how each particular data element looks. Is "Product Code" a character or numeral? Is "Start Date" a date or is it really a datetime stamp?

As explained in Chapter 5, data types are important in statistical analysis because they dictate which statistics and statistical procedures can be applied to which data elements. If you try to take the average value of a character field like "First Name," you're going to get an error message every time.

Typically, this type of metadata is pretty accurate. It is generally stored by the system that holds the data and can be generated automatically. Verifying the formats is generally pretty straightforward. Such verification is essentially a by-product of the validation of data ranges discussed in the following section. But there are instances where it can be a bit more difficult.

We've seen one such scenario more times than we care to recall. It happens sometimes that when a system is first designed, the development team tries to put some flexibility into the data structures to accommodate future enhancements. Sometimes they just add a bunch of empty (and wide) alphanumeric data columns onto the end of each record. These auxiliary columns are initially not used for anything.

Analysts will always err on the side of asking for more data rather than less — frequently, *all* data rather than *some*. This fact, combined with the need to get the data quickly, sometimes results in a *data dump*. This dump generally includes the auxiliary columns. In these cases, the metadata tells you something like "Fields 1–11" are formatted as "200 alphanumeric characters."

Such information is practically useless. To make sense of a data field like this, you pretty much have to get your hands dirty. There's not a lot you can do except page through a few dozen records and try to make an informed guess about what's actually in the field. In most cases, these fields tend to be empty. But not always. The good news is that if the field is actually being used, you should be able to find a programmer somewhere who knows what it's being used for.

Typecasting your data

One of the most critical steps in performing a statistical analysis is making sure that your data is what it purports to be. Statistical procedures will invariably crash if you don't provide them with valid information about data formats. But these procedures are largely blind to problems with the validity of the data.

Understanding how a data field is formatted is not enough. Before turning a dataset over to a statistical procedure, you need to understand what the data actually is in each of the fields you are using.

As discussed in Chapter 5, most data falls into one of four categories: nominal, ordinal, interval, and ratio. The data type determines what sort of statistics and statistical procedures can be applied to particular data fields. You can't take an average of a field like "Last Name," for example.

Confusing data types with data formats is easy (and far too common). Knowing whether a data field is a character, integer, or continuous does not tell you the data type.

As mentioned earlier, character fields are sometimes used as placeholders for data that might be captured in future releases of a system. There is nothing to prevent such a field from being used to capture monetary or other numeric data.

The most common data-type mistake involves assuming that a numeric field, particularly an integer-valued field, actually contains numeric *ordinal* data. It is extremely common for companies to use numeric codes (*nominal* data) to represent products, regions, stores, and various other entities.

Airline flight codes are one example. Census regions are another. Even credit card and Social Security numbers are typically stored as integers. But all of these entities are merely identifiers. They are *nominal* variables. The *average credit card number* in a bank's portfolio is a meaningless statistic.

Being Careful with Dates

We very rarely run across a dataset that does not include dates. Purchase dates, birthdates, update dates, quote dates, and the list goes on. In almost every context, some sort of date is required to get a full picture of the situation you are trying to analyze.

Dealing with dates can be a bit tricky, partly because of the variety of ways to store them. But also, depending on what you're trying to do, you may only need part of the date. This section points out a few common situations to look out for.

Dealing with datetime formats

For starters, most database management systems have an extremely precise way of storing dates internally: They use a *datetime.* This is exactly what it sounds like: a mashup of the date and the time. For example, a common format looks like this:

$$2014-11-24\ 14:25:44$$

That means 25 minutes and 44 seconds past 2 p.m. on November 24, 2014.

The seemingly excessive detail here is rarely fully utilized. By far the most common user of the full detail is the database management system itself. It is a common practice for databases to put a datetime stamp on every record to indicate when the record was created and when it was last updated. The New

York Stock Exchange systems actually keep track of trade time stamps to even greater precision.

For most analytic applications, however, this is more detail than you want.

If you are analyzing a stock's closing price over time, you won't be interested in more than just the day or maybe the month associated with each closing price. If you are doing a demographic analysis of age distributions, the year of birth may be all that's relevant.

Birthdates provide a good example of something that you may encounter with datetime data. Even though data may be stored in a datetime field, it may be the case that only part of the field is really being used. Birthdates typically have the time portion defaulted to 00:00:00 for every record.

Luckily, both database systems and analytic software have built-in functions that allow you to extract only the portion of the datetime that is relevant to you. You can choose to extract only the date part, only the month and year, only the year, and so forth. And in fact, this is often done for you before you ever see the data.

Taking geography into account

In the brave new world of the global economy, you will likely encounter data that has been collected from many different locations. Anyone who has ever tried to schedule an international conference call is well aware of the logistics involved in dealing with multiple time zones. More and more common nowadays are post-midnight conference calls with India.

One typical big data example involves supply chain management. *Supply chain management* is the ongoing process of trying to manage raw materials, inventories, distribution, and any other relevant aspect of a company's business. It's how Walmart keeps shelves stocked, how UPS keeps track of packages, and how Amazon manages to deliver almost anything imaginable almost anywhere.

In these examples, the analysis that underlies supply chain management needs to take into account that data is coming from different time zones. When faced with situations like this, datetime data must be dealt with carefully.

Suppose a package is shipped from California at 10 a.m. on Wednesday and is delivered to its final destination in New York on Thursday at 10 a.m. If you are interested in analyzing delivery times, you need to take into account the time zone change. In this example, the delivery time is actually 21 hours, not 24.

When dealing with datetime data collected from different time zones, you can't simply compare different data points based on the raw data. You need to first make sure that all datetimes are represented in a common time zone. Which time zone you use is somewhat arbitrary, so long as all the data points are using the same one.

There is one other geographically — or, to be more accurate, culturally — related fact that you need to be aware of. Not all countries represent dates in the same way. The U.S. is actually somewhat unique in representing dates as month/day/year. Canada and most of Europe prefer to use the convention day/month/year. You may also run across variations beginning with the year.

How your software thinks about dates

Dates are used in a variety of ways in data analysis. Sometimes, as with stock price analysis, their primary function is to put the observations in order from earliest to latest. But in other cases, they are used to measure time intervals.

In engineering, particularly in quality control applications, a key statistic is *mean time to failure*. This is simply the average life span of a part or product. For long-lived products, like car parts and light bulbs, this calculation requires the comparison of dates.

On the face of it, August 15, 2013 minus January 1, 2010 doesn't make much sense mathematically. We all know what is meant by this, but it takes some thinking to get the answer. For this reason, many statistical packages, when confronted with dates, immediately convert them into a number in order to facilitate comparisons. They do this by picking some starting point and calculating the number of days between that starting point and the date that is being converted.

For example, one large statistical software maker, SAS, uses the date January 1, 1960 as its starting point. This date has the value 0. It stores every date as the number of days it is away from this starting point. Thus, SAS thinks of January 1, 1961 as 366 (remember, 1960 was a leap year, and January 1 is day 0, not day 1). The starting point is arbitrary and different software makers use different starting points, but the idea is the same.

One odd consequence of this convention is that if you look at the raw data, not only are all the dates integers, but they don't even have to be positive integers. In the SAS example, January 1, 1959 would be represented as −365.

In any case, this way of handling dates facilitates calculations. By converting the date to a number on input, the system avoids having to jump through hoops every time a calculation involving that date is performed.

Does the Data Make Sense?

It's important to verify that the data in each field actually makes sense. Once you have determined the format and data type for each field, you need to dig into each field to make sure it doesn't contain inconsistencies. The basic idea is to formulate some basic properties that the data should exhibit and verify that these are indeed reflected in the data.

The best way to do this is to graph your data. Depending on whether the data is discrete or continuous, you will use slightly different techniques to do this. Chapter 12 is dedicated to graphing techniques. This section focuses on what you're looking for.

Checking discrete data

One common situation is for a data field to have a predetermined list of values that populate the field. This can happen with *nominal* data. Product codes and flight numbers are good examples. In these cases, it is generally possible for you to lay your hands on an accurate list of valid codes.

When you have a relatively small list of valid codes, a simple test will give you some confidence in your data. Simply generate a list of all the codes that appear in your dataset and compare it to the list of valid codes.

If you have codes in your dataset that do not appear on the list of valid codes, this is a problem. Either your list is not up to date or your data is invalid. You'll need to track down the source of the issue before you can use this field in your analysis.

Conversely, if you have more valid codes in your list than appear in your dataset, this could be a problem. Frequently this happens because old codes have been decommissioned but not deleted from the valid code list. They were valid at one time. But it could also mean that your dataset is not complete. If you're analyzing product usage and your dataset does not contain any codes representing your most popular product, you have a problem. In that case, you need to go back to the source of your dataset to investigate. The validity of the entire dataset is now in question.

This approach works well for some ordinal data as well. You can use it to verify that the data is in the appropriate range. For example, many survey datasets contain answers to question like "On a scale of 1 to 10, rate . . . ." These data fields should not contain 11s.

For ordinal data that is not in a pre-defined range, you should still get a list of the values that appear in your dataset. For example, if you're analyzing personal automobile ownership, your data may not have a predetermined maximum in the "Number of Cars Owned" field. And indeed, there are some car junkies out there who might have 10 or 20 cars. But if you see 200, you may want to scratch your head a little bit over what's going on. There is a more detailed description of this situation in Chapter 10, which discusses *outliers*.

For some nominal data, this approach is not practical, or even necessary. In cases where you have some sort of account number, this number is typically not relevant to your analysis. Instead, it's being used simply to identify each record uniquely. In that case, you might focus on verifying that the numbers have the correct length. Credit card numbers all have 16 digits, for example.

Account numbers and other identifier type fields are useful in validating the quality of your dataset. In particular, you can use them to identify problems involving duplicate records. Duplicates are problematic for statistical analysis. The more times a record appears, the more importance it is given in the outcome of the analysis. In other words, the duplicate records will be over-counted, skewing your results.

Checking continuous data

Examples involving financial data occur throughout this book. These examples involve stock prices, account balances, and interest rates. The statistical techniques in this book always treat this data as continuous. When dealing with continuous data, it is again impractical (not to mention uninformative) to try to look at every value in the dataset.

When evaluating the quality of continuous data, the best place to start is by simply looking at the distribution graphically. See Chapter 12 for a discussion of graphical techniques.

Looking at the width of the distribution

You can uncover a couple of problems by simply looking at the largest and smallest values in a data field. That's a quick way to diagnose an outlier problem. If you find the range to be unreasonably large, it's an indication that your dataset might contain records that are out of whack with the rest of the data. This can cause your analysis to be skewed and misleading. Turn to Chapter 10 for further discussion of outliers.

But what does *unreasonably large* (or small) mean? This is where your task becomes part art, part science. The interpretation of "unreasonable" depends on the context. It requires some practical knowledge of the situation you are trying to analyze.

For example, if you are analyzing mortgage interest rates for the 5-year period from 2009 to 2014, you might raise your eyebrows if you are seeing 15-percent interest rates. On the other hand, if you are looking at the late 1970s, this wouldn't be the least bit unusual. In this case, recognizing that 15 percent is not a mortgage rate you would expect to see in a recent dataset depends on your knowing something about the history of mortgage rates.

But common sense is a good practical guide. You wouldn't expect to see negative values in a data field purporting to contain a stock price. A negative balance in a savings account might give you pause. An interest rate field probably wouldn't contain negative values, but a rate of return field might.

In evaluating the quality of any data field, it is not sufficient to simply run statistical tests. It is important have a clear understanding of what the data is purported to represent. The data is supposed to represent some fact about the real world. Your job is to verify that this is happening.

Looking at the shape of the distribution

Having said that, there are some purely technical reasons to be aware of the shape of the distribution of your data. A variety of statistical modeling techniques are discussed in the chapters in Parts III and IV of this book. Many of these processes are fairly picky about the shape of data distributions.

For example, some modeling techniques operate under the assumption that continuous variables are normally distributed. In other words, they expect all the distributions to follow a nice bell-shaped pattern. This means distributions that are skewed to the right or left can adversely affect the usefulness of the model.

But if a distribution does not meet the assumptions of a statistical procedure, all is not necessarily lost. In some cases, it is possible to perform some simple transformations of the data to give the distribution an acceptable shape.

There are several examples of such data transformation later in this book. Chapters 13 through 16 in particular discuss some common approaches friendly to statistical procedures. Detailed discussion of these transformations can be found in those chapters.

For now, just know that the shape of data distributions is important. Different statistical techniques have different requirements. Problematic distributions need to be identified and corrected up front, *before* the data is fed to a statistical modeling procedure.

Frequently Encountered Data Headaches

There are several common issues that seemingly always crop up in large databases. This chapter has already touched on one of these issues — *outliers*. Indeed, that subject has a chapter of its own. The issues discussed in this section, like outliers, are somewhat insidious because they may not cause a fatal error when you run a modeling procedure on your data. The procedure may not even notice them. But when these issues are present, they can adversely affect the quality of your output.

Missing values

One of the most frequent and messiest data problems to deal with is missing data. Files can be incomplete because records were dropped or a storage device filled up. Or certain data fields might contain no data for some records. The first of these problems can be diagnosed by simply verifying record counts for files. The second problem is more difficult to deal with.

Chapter 9 is dedicated to the subject of missing data, so you only get a brief introduction to the topic here. To put it in simple terms, when you find a field containing missing values, you have two choices:

✔ Ignore it.
✔ Stick something in the field.

Ignoring the problem

In some cases, you may simply find a single field with a large number of missing values. If so, the easiest thing to do is just ignore the field. Don't include it in your analysis.

Another way to ignore the problem is to ignore the record. Simply delete the record containing the missing data. This may make sense if there are only a few rogue records. But if there are multiple data fields containing significant numbers of missing values, this approach may shrink your record count to an unacceptable level.

Another thing to look out for before simply deleting records is any sign of a pattern. For example, suppose you are analyzing a dataset related to credit card balances nationwide. You may well find a whole bunch of records showing $0.00 balances (perhaps around half the records). This is not in itself an indication of missing data. However, if all the records from, say, California are showing $0.00 balances, that indicates a potential missing values problem. And it's not one that would be usefully solved by deleting all the records from the largest state in the country. In this case, it is probably a systems issue and indicates that a new file should be created.

In general, deleting records is an easy, but not ideal, solution to missing-value problems. If the problem is relatively small and there is no discernible pattern to the omissions, then it may be okay to jettison the offending records and move on. But frequently a more highbrow approach is warranted.

Filling in the missing data

Filling in the missing data amounts to making an educated guess about what would have been in that field. There are good and bad ways to do this. One simple (but bad) approach is to replace the missing values with the average of the non-missing ones. In non-numeric fields, you might be tempted to populate the missing records with the most common value in the other records (the mode).

These approaches are, unfortunately, still frequently used in some business applications. But they are widely regarded by statisticians as bad ideas. For one thing, the whole point of doing statistical analysis is to find data that differentiates one result from another. By replacing all the missing records with the same value, you haven't differentiated anything.

The more highbrow approach is to try to find a way to predict in a meaningful way what value should be filled in on each record that is missing a value. This involves looking at the complete records and trying to find clues as to what the missing value might be.

The statistics behind this approach are examined in Chapter 9, so the discussion here is restricted to an illustrative example. Suppose you are analyzing a demographic file to predict likely purchasers of one of your products. In that file you have, among other fields, information on marital status, number of children, and number of automobiles. For some reason, the number of autos field is missing in one-third of the records.

By analyzing the other two fields — marital status and number of children — you may discover some patterns. Single people tend to have one car. Married people with no children tend to have two cars. Married people with more than one child might be more likely to have three cars. In this way, you can guess at the missing values in a way that actually differentiates the records. More on this approach to come.

There is a general term in statistics and data processing that refers to questionable data. The term *noisy* is used to describe data that is unreliable, corrupt, or otherwise less than pristine. Missing data is but one example of this. A detailed description of techniques for cleaning up noisy data in general is beyond the scope of this book. In fact, this is an active area of research in statistical theory. The fact that all noise is not as easy to spot as missing values makes it troublesome to deal with.

Duplicate records

Data is stored in different ways in different systems. In particular, what makes an individual record unique is different for different systems. An investment account summary is attached to an account number. A portfolio summary might be stored at an individual or household level. And the trading histories of all those accounts are stored at the individual transaction level.

So it's no surprise that when collecting and consolidating data from various sources, it's possible that duplicates pop up. For this reason, it's important to be clear about what is supposed to differentiate unique records in your data file. For example, if it's a transaction level file, then account numbers and household IDs will be duplicated. As long as you understand this and are doing a transaction level analysis, you will be fine.

But if you are interested in using this data to analyze the number of accounts held by each household, you will run into problems. The households that trade more frequently will have more records than those that don't trade very much. You need to have a file at the account level.

Removing duplicate records is not particularly difficult. Most statistical packages and database systems have built-in commands that group records together. (In fact, in the database language SQL, this command is called Group By.)

Other Common Data Transformations

It is frequently desirable to standardize data before performing advanced statistical procedures. Standardizing can make the data more manageable. It can also improve the accuracy of models as well as make them a little easier to interpret. This section discusses some common ways of standardizing data.

Percentiles

Anyone who's ever taken a standardized test like the SAT is familiar with seeing the scores reported as a percentile. If you're in the 90th percentile, it means that 90 percent of the scores on the test are at or below your score.

There is nothing particularly special about percentiles. You can do the same sort of thing for any number of *quantiles*. Instead of using 100 percentiles, you can split the data up into any number of buckets you wish. Two common options are 4 *(quartiles)* and 10 *(deciles)*.

When you are confronted with data that is widely spread out, it is sometimes helpful to convert it into quantiles. For example, income data in the U.S. has a very wide distribution. If you feed that raw data into a model, much of the important variation at the lower end of the income scale (say below $50,000) gets swallowed up by the huge variation that happens among the top 1 percent (say $500,000 up to $100,000,000 or more). $10,000 makes a lot bigger difference to someone in the 10th percentile than to someone in the 90th. That variation at the lower end can be more effectively captured if the incomes are reported in percentiles instead of raw dollars.

This sort of conversion is best used when you are concerned mainly with variation. If you're using the data to see who is capable of affording a Ferrari, then this sort of transformation may not be very useful.

Standard scores

Another common transformation is to convert data to *standard scores*. Essentially, you calculate the mean and standard deviation of a distribution. The standard score is calculated by subtracting the mean from each observation and dividing by the standard deviation. In most cases, the vast majority of standard scores will lie somewhere between −3 and 3.

In the context of predictive models, this is sometimes a beneficial thing to do. When you have variables in your data that are on wildly different scales, it's a way of standardizing the variation so that it's given equal weight for each variable. For example, suppose you have interest rate data ranging from 0 to 10 percent combined with investment account data with balances ranging into the hundreds of millions.

In this case, some procedures can get lost in the larger range of the balance data and downplay the importance of the interest data. At the very least, the resulting model can end up with some very large and unwieldy parameters that can be difficult to interpret. In these cases, it can be helpful to convert your input variables into standard scores. This puts all variables on an equal footing with respect to their variation. In the resulting model, it will then be much easier to see what variables are really contributing to the predicted outcomes.

Dummy variables

Statistical procedures, for the most part, require numeric variables. They don't deal well (or at all) with "Vanilla," "Chocolate," and "Strawberry." So if you're trying to analyze what ice-cream flavors are associated with the biggest food orders at your ice-cream shop, you need to apply a common trick. *Dummy variables* are numeric variables that indicate whether a certain condition is true or false. Typically they are assigned a value of 0 if a condition is false and a 1 if that condition is true.

For the ice-cream example, you might create three dummy variables called "Vanilla," "Chocolate," and "Strawberry." For every order, if the order contains vanilla ice-cream, you set the "Vanilla" variable to 1. Otherwise it is set to 0. Similarly, the other two variables are each assigned 0s and 1s. In this way, you've converted your text to logical (sometimes called *Boolean,* after a famous logician) variables that can be handled by appropriate statistical procedures.

There is one subtlety lurking here: How you choose your dummy variables depends on the nature of the data. In particular, how many dummy variables you create depends on whether at least one of the conditions is always met.

The ice-cream example was leaving open the possibility that an order could contain no ice-cream at all. Thus all three variables could potentially be 0. However, if you have a situation where at least one of the conditions needs to be met, you actually need to include one fewer variable. Take, for example, data that contains the day of the week that each record was created. If you create six dummy variables for Sunday through Friday, then a seventh one is not necessary. Saturday is indicated by all six of the variables being 0.

There is a little more going on here than the seventh variable being unnecessary. Including it actually introduces a redundancy in the data that can be problematic for predictive models.

As you can see, there is a lot more to doing statistical analysis than just feeding a dataset to a statistical procedure. The quality and usefulness of your results will be greatly enhanced if you take the time and effort to heed the wisdom of "Garbage in, garbage out!"

Chapter 7

Figuring the Format: Important Computer File Formats

In This Chapter

▶ Understanding different computer data formats, such as .csv, .xlsx, and .html

▶ Getting acquainted with data processing in Excel

▶ Learning the basics of web data formats

*T*here are practically as many data formats as there are systems that create and store data. When doing statistical analysis on that data, you very frequently, if not always, will be using PC-based software. Obviously, then, your data will end up on your PC. In order to help you manage this data, this chapter looks at some key features of several important formats for storing and analyzing data using spreadsheets and database programs.

Spreadsheet Formats

You can save spreadsheet programs such as Microsoft Excel with several different formats, including the following:

- Comma-separated variable (.csv)
- Tab-delimited text (.txt)
- Space-delimited (.prn)
- Excel Workbook (.xlsx)
- Excel macro-enabled workbook (.xlsm)
- Excel Binary Workbook (.xlsb)
- XML (Extensible Markup Language) (.xml)
- HTML (Hypertext Markup Language) (.html)

Comma-separated variables (.csv)

A commonly used structure for storing data is the *record*. Each record consists of a series of *fields,* and each field in a record contains a single value. Each field is separated, not very surprisingly, by a comma. These fields become the variables that you analyze with statistical tools. Table 7-1 shows the record structure of a company's employee database.

Table 7-1	Structure of Records in an Employee Database			
Last Name	First Name	Social Security Number	Age	Department

The fields in this record are: Last Name, First Name, Social Security Number, Age, and Department.

With the comma-separated variables (CSV) format, records consist of *plain text.* That is, the text doesn't contain any special formatting characters. In a CSV file, a record from the customer database in Table 7-1 would appear as follows:

Smith, John, 111-11-1111, 62, Accounting

With CSV, each record is separated from the other records by placing a "newline" character or "end of line" character at the end of the record.

You can easily identify a comma-separated variables file; it usually has a file extension of .csv. For example, a file named MyRecords.csv contains data stored in the comma-separated variables format.

CSV uses a special character set to represent data, but note that different versions of CSV use different character sets. ASCII (short for American Standard Code for Information Interchange) and Unicode are two of most commonly used character sets.

ASCII

ASCII is a system for converting computer bits into a character or a number. For example, ASCII translates 1000001 into capital *A* and 1100001 into lowercase *a.*

A *bit* is a unit of computer storage that can only assume a value of 0 or 1. A collection of bits is known as a *byte* (commonly 8 bits). A single byte can represent a letter, number, or other character.

ASCII was first based on seven bits, so it could only represent 128 (2^7) different characters and numbers. After being developed in the 1960s, ASCII was later extended to eight bits. This made it possible to represent up to 256 (2^8) different characters and numbers. ASCII was widely used on the Internet until the late 2000s, when it was overtaken by Unicode.

Unicode

Unicode is a more comprehensive system than ASCII. It was created in the late 1980s and has superseded the use of ASCII. Unicode may be implemented with different encoding systems; two of these are known as UTF-8 and UTF-16. (UTF stands for Unicode Transformation Format.)

UTF-8 consists of the ASCII system plus a large number of additional characters and is now widely used in applications and for web pages on the Internet. Unlike ASCII, UTF characters often consist of more than byte. With UTF-8, each ASCII character is represented with a single byte, but the additional characters are represented with anywhere from two to four bytes. UTF-16 is based on 16-bit bytes, and isn't compatible with ASCII.

One of the advantages of Unicode compared with ASCII is that ASCII is restricted to Western European languages, whereas Unicode can be extended to languages that contain different types of symbols, such as Greek, Cyrillic, Hebrew, and so on.

One of the reasons for the popularity of the CSV format is that it is compatible with a large number of applications. It can also be easily converted into other formats, such as a Microsoft Excel spreadsheet.

Text

Text is a file format that contains primarily plain text, but may contain other characters such as numbers and symbols as well. Probably most desktop applications can create text files. Text editors, including the simple Microsoft Notepad, create nothing but text files.

Within a text file, values are typically *delimited* (that is, separated) by either a tab character or by a space. The appropriate file extensions for these formats are as follows:

- ✔ Tab-delimited files (.txt)
- ✔ Space-delimited files (.prn)

Text files don't store any formatting information, which makes them compatible with a wide variety of applications. They can be imported into a spreadsheet program such as Microsoft Excel and also converted into other formats. One of the drawbacks to text files is that storage is inefficient; in other words, text files require more storage space than many other formats.

Microsoft Excel

Unlike CSV-formatted files, Excel files store data in binary format. A *binary* file may contain text, but it may also contain other types of information, such as formatting, sound, pictures, video, and more.

Microsoft Excel can store spreadsheets in many different formats. Here are three of the most important formats:

- ✔ Excel workbook (.xlsx)
- ✔ Excel macro-enabled workbook (.xlsm)
- ✔ Binary workbook (.xlsb)

Excel workbook (.xlsx)

.xlsx is the standard format for Excel spreadsheets, starting with Excel 2007. Previously, the standard format was .xls. The .xls format was split up into three new formats: .xlsx, .xslm, and .xlsb.

The .xlsx format is based on the Extensible Markup Language (XML), and is intended to increase the compatibility of Excel spreadsheets with other applications. One advantage of the .xlsx format is that it requires less storage space than the older .xls format, due to file compression. Another huge advantage to .xlsx is that it's much easier to recover corrupted files than with .xls.

The .xlsx format doesn't allow user-defined programs known as *macros*. You must save spreadsheets containing macros in a different format — .xlsm format (Excel macro-enabled workbook). The advantage of this setup is that the user can open an .xlsx spreadsheet without having to worry about whether there's a macro attached to the file. This is an important security measure, because viruses can be spread with macros.

Another advantage of the .xlsx format is that Excel spreadsheets now contain more data than in the past. The .xls format only allowed for a maximum of 65,536 rows and 256 columns; in case those weren't enough for you, the .xlsx format allows for 1,048,576 rows and 16,384 columns. .xlsx also provides superior handling of tables, making it easier to add new information to an existing table. This is clearly a huge advantage for big data!

Excel macro-enabled workbooks (.xlsm)

Microsoft Excel contains a programming language called Visual Basic for Applications (VBA). The commands in VBA closely resemble English. You can use VBA as a traditional programming language to do many tasks, such as creating new add-in functions for Excel. You can also use VBA to produce *macros,* which are small programs used to automate routine tasks in Excel.

Creating new Excel functions

You can use VBA to extend the capabilities of Excel by creating new functions. Excel then executes these functions as if they were part of the original package.

For example, suppose you find yourself frequently computing cube roots in Excel. There's no built-in cube root function, so you must enter a formula each time you want to do this calculation. As an alternative, you could create a VBA function called "CubeRoot" to extend Excel's capabilities with the following set of keystrokes: `=(A1)^(1/3)`.

Suppose you need to know the cube root of 8, which is stored in cell A1 on your spreadsheet. By typing the command `=CubeRoot(A1)` in another cell, you cause Excel to produce the correct result of 2.

Automating tasks with macros

You can write VBA code to create macros by enabling the Developer tab in Excel. In Excel 2010 or later, click File > Options. In the Options dialog box, choose Customize Ribbon. In the Customize Ribbon dialog box, choose Main Tabs; this produces a list of tabs such as Home, Insert, and so forth. Most likely, all of them are selected except for Developer.

Click the box for Developer and then click OK. You'll notice that the Developer tab now appears next to the View tab in the menu bar. From the Developer tab, click Visual Basic to write your own code or click Macros to execute a macro.

You can also use the Macro Recorder to create macros that automate tasks. From the Developer tab, click the Record Macro button. This produces a dialog box in which you can provide the macro's name, an optional short-cut key, and a specification of where the macro is to be stored. Once the recorder starts, it stores all the keystrokes that follow until you turn it off.

You can use the macro recorder for many purposes:

- Opening files
- Creating new documents
- Updating existing documents
- Saving files
- Formatting cells

Whenever you create VBA code or macros in Excel, you must save the spreadsheet as an Excel macro-enabled workbook (.xlsm) in order to save the code.

Spreadsheets are very powerful tools that can be customized to perform almost any data-processing task (including statistical procedures). For more on what you can do in Excel spreadsheets, check out *Excel VBA Programming For Dummies* (Wiley, 2013) by John Walkenbach.

The standard Excel workbook format (.xlsx) doesn't support VBA code. In fact, the standard workbook removes VBA code when it's saved.

Binary workbook (.xlsb)

The binary workbook format (.xlsb) is very similar to the standard Excel workbook format (.xlsx). The binary workbook format is designed to speed up the process of opening and saving files. This format is particularly advantageous with extremely large files.

Unlike the .xlsx format, .xlsb format *does* support macros. The main drawback to using .xlsb is that it's typically not compatible with other applications.

Web formats

The Web is by definition a distributed computing system. Data is spread out, quite literally, all over the world. It is processed and transferred by a seemingly infinite number of machines. For this reason, web data is standardized in a way that is independent of the particular machine that's processing it. This section gives a brief description of the way this is typically done.

XML

Extensible Markup Language (XML) is a language that lets users define the meaning of elements in a document, spreadsheet, or other type of application file through the use of *tags*. You define tags in a separate file known as a *schema*.

As an example, the following XML code represents a single record containing the statistics of a baseball player:

```
<row>
<Year>2014</Year>
<Homeruns>39</Homeruns>
<RBI>142</RBI>
<Average>0.344</Average>
</row>
```

Using tags makes XML a very flexible and powerful language. XML is also independent of the operating system being used (for example, it works in Windows and Macintosh OS X).

You can save Excel .xml files, providing you define a schema. These .xml files can be opened in many other applications that support XML. By converting files to an .xml format, you allow your data to be exported to other applications that don't accept Excel files.

HTML

Hypertext Markup Language (HTML) is a language used to create web pages on the Internet. The HTML language consists of elements identified by *tags*. For example, to start a new paragraph, you use the <p> tag to indicate the beginning. At the end of the paragraph, you use the </p> tag to indicate the end. You can save Excel files as HTML files that you can publish directly to the Internet.

You can view the HTML code used to create a web page by right-clicking the menu bar in your browser and choosing View ⇨ Source (Internet Explorer) or clicking Tools ⇨ Web Developer ⇨ Page Source (Mozilla Firefox).

Database Formats

A *database management system* is software that is specifically designed to store and manage data. There are many different such systems. Two of the most accessible (meaning inexpensive) are Microsoft Access and MySQL. Each database type has its own unique format. For example, Microsoft Access uses a format known as .accdb.

Microsoft Access (.accdb)

Microsoft Access is a relational database. A relational database organizes data into tables.

The Microsoft Access format .accdb was introduced with Access 2007. It's an upgrade from the previous format (.mdb). The .accdb format offers several new features. One of these features is the ability to publish a database directly to the Internet. Another is the ability to store more than one value in a record.

For example, suppose you create a database to keep track of a company's inventory. One of the fields is `color`; for an item that has more than one color, the field can store each of them.

Another new feature is the attachment data type. With this data type, it's possible to store a wide variety of file types in a database without requiring a large amount of storage.

MySQL (.frm)

MySQL is an open source relational database system. Structured Query Language (SQL) is a programming language specifically designed for accessing information from a database. MySQL uses a format known as .frm.

Oracle first developed MySQL in 1995. In addition to open source MySQL, you can buy more advanced versions from Oracle and other vendors.

In addition to being free, one of the great advantages of MySQL is its ability to take advantage of cloud computing, where processing is done over the Internet. Cloud computing has several advantages:

- ✔ It can be less expensive than traditional computing.
- ✔ It provides virtually unlimited storage capabilities.
- ✔ Backing up and recovering files are easier than with traditional computing.
- ✔ The latest software is always available.
- ✔ Data can be accessed from any computer and any location.

One of the potential drawbacks to cloud computing is that data may be more vulnerable to hackers.

There is nothing more frustrating than getting a dataset that you're eager to analyze and not being able to open it. Or, just as bad, opening it and seeing unintelligible gibberish. In this chapter, we have given you at least a hint of some of the things that could be going on in such cases. We have by no means covered the topic of data processing in detail. But hopefully we've given you a sense of where to begin when faced with a data formatting challenge.

Chapter 8

Checking Assumptions: Testing for Normality

In This Chapter

▶ Putting in place a goodness of fit test

▶ Implementing a Jarque-Bera test

*A*ll areas of statistical analysis make assumptions about the data being studied. To ensure the validity of any statistical tests performed on a dataset, determining whether the assumptions made are actually correct is essential.

One of the most common assumptions in business disciplines such as economics, finance, marketing, and so forth is that a variable is *normally distributed*. In particular, rates of return to financial assets are often assumed to be normally distributed. If this assumption is incorrect, the results of any statistical tests will be questionable.

Many tests are specifically designed to determine whether a dataset follows the normal distribution. This chapter examines two of them in detail: the goodness of fit test and the Jarque-Bera test.

Goodness of fit test

You use a *goodness of fit test* to test the hypothesis that a population conforms to a specified probability distribution. For example, you can use a goodness of fit test to determine whether a population is normally distributed. The goodness of fit test is based on the *chi-square* distribution.

Chapter 5 introduces the notion of hypothesis testing. There we test hypotheses about a population mean, and the test statistic is normally distributed.

But hypothesis tests can be implemented in a variety of situations. And depending on the situation, different test statistics are appropriate. In the case of the goodness of fit test, the test statistic follows a chi-square (pronounced "ki-square") distribution, described next.

The chi-square distribution

A chi-square random variable is composed of a sum of *independent, standard normal* random variables that have been squared. This is shown in the following equation:

$$\chi_v^2 = Z_1^2 + Z_2^2 + Z_3^2 + \ldots + Z_v^2$$

In this equation, v standard normal random variables are combined; v is the Greek letter *nu* and is often used in statistics to represent the number of degrees of freedom of a probability distribution.

The chi-square distribution has several key properties:

- ✔ It is positively skewed.
- ✔ Its shape depends on its *degrees of freedom (df)*. This is the number of standard normal variables that appear in the sum.
- ✔ It is undefined for negative values. This is because all the component normal variables have been squared, making them positive.

In the section in Chapter 5 on hypothesis testing, we distinguish between one- and two-tailed tests. It makes sense to talk about a two-tailed test in the context of the normal distribution. The normal distribution actually has two tails. It goes on forever in both the positive and negative directions.

But the chi-square distribution is different. Because of the positive skew and the fact that the variable is positive, chi-square distributions only have a positive tail. So all chi-square based hypothesis tests are implemented as one-tailed tests.

In what follows, we show you how to use the chi-square distribution to test whether you can reasonably assume that a variable is normally distributed.

The null and alternative hypotheses

In this case, the null and alternative hypotheses would be:

H_0: The population is normally distributed.

H_1: The population is not normally distributed.

The level of significance

The level of significance refers to probability of committing a Type I error, and is usually set equal to 0.05 in financial applications. As Chapter 5 mentions, a Type I error consists of rejecting the null hypothesis when it's actually true.

Computing the test statistic

In order to test the hypothesis that a population follows a specified distribution, you must construct a table that shows the *observed* distribution of the sample elements, organized into categories, and the *expected* distribution of the sample elements. The table shows the number of elements that would fall into each category if the assumed distribution is true. The closer the observed values are to the expected values, the more likely it is that the population follows the assumed distribution.

The number of categories corresponds to the degrees of freedom (df) of the chi-square distribution you will use. Choosing the ideal number of categories is something of an art. In essence, you want to choose few enough so that the categories aren't too small. The test isn't meaningful if you have a bunch of ranges that don't contain any data.

For example, suppose an investor wants to determine whether the returns to his or her portfolio are normally distributed with a mean of 5 percent and a standard deviation of 10 percent. The portfolio consists of 100 stocks. Table 8-1 shows the observed frequencies for the returns.

Table 8-1	Portfolio Returns			
Returns	*−15% to −5%*	*−5% to +5%*	*+5% to +15%*	*+15% to +25%*
Observed frequency	22	29	37	12

The null and alternative hypotheses are as follows:

H_0: The population is normally distributed with mean = 5 percent and standard deviation = 10 percent.

H_1: The population is *not* normally distributed with mean = 5 percent and standard deviation = 10 percent. This amounts to saying the null hypothesis is false.

We choose the level of significance to be 5 percent.

As indicated earlier, a goodness of fit test is always implemented as a *right-tailed* test.

The probability of a stock's return being between –15 percent and –5 percent when the population returns are normally distributed with a mean of 5 percent and a standard deviation of 10 percent can be computed as follows. First, convert these probabilities into *standard normal form.* You do this with the following equation:

$$Z = \frac{X - \mu}{\sigma} = \frac{X - 0.05}{0.10}$$

The key terms in this equation are as follows:

Z is a *standard normal* random variable; its mean is 0 and its standard deviation is 1.

X is a normal random variable; its mean is .05 and its standard deviation is *.10.*

The probability that X is between –15 percent and –5 percent can be *standardized* as follows:

$$P(-0.15 \leq X \leq -0.05) = P\left(\left(\frac{-0.15 - 0.05}{0.10}\right) \leq Z \leq \left(\frac{-0.05 - 0.05}{0.10}\right)\right)$$
$$= P(-2.00 \leq Z \leq -1.00)$$
$$= P(Z \leq -1.00) - P(Z \leq -2.00)$$

$P(Z \leq -1.00)$ and $P(Z \leq -2.00)$ can be found with a statistical calculator or standard normal tables. Also, you can find them with the Excel function *normsdist.* For example:

$$P(Z \leq -1.00) = \text{normsdist}(-1.00) = 0.1587$$

$$P(Z \leq -2.00) = \text{normsdist}(-2.00) = 0.0228$$

Therefore,

$$P(Z \le -1.00) - P(Z \le -2.00) = 0.1587 - 0.0228 = 0.1359$$

This shows that if the returns to the portfolio are normally distributed with a mean of 5 percent and a standard deviation of 10 percent, then 13.59 percent of the returns should be between –15 percent and –5 percent.

The remaining probabilities are computed as follows. The probability that X is between –5 percent and +5 percent is

$$P(-0.05 \le X \le 0.05) = P\left(\left(\frac{-0.05 - 0.05}{0.10}\right) \le Z \le \left(\frac{0.05 - 0.05}{0.10}\right)\right)$$
$$= P(-1.00 \le Z \le 0.00)$$
$$= P(Z \le 0.00) - P(Z \le -1.00)$$
$$= 0.5000 - 0.1587 = 0.3413$$

This shows that if the returns to the portfolio are normally distributed with a mean of 5 percent and a standard deviation of 10 percent, then 34.13 percent of the returns should be between –5 percent and 0 percent.

The probability that X is between +5 percent and +15 percent is

$$P(0.05 \le X \le 0.15) = P\left(\left(\frac{0.05 - 0.05}{0.10}\right) \le Z \le \left(\frac{0.15 - 0.05}{0.10}\right)\right)$$
$$= P(0.00 \le Z \le 1.00)$$
$$= P(Z \le 1.00) - P(Z \le 0.00)$$
$$= 0.8413 - 0.5000$$
$$= 0.3413$$

This shows that if the returns to the portfolio are normally distributed with a mean of 5 percent and a standard deviation of 10 percent, then 34.13 percent of the returns should be between +5 percent and +15 percent.

The probability that X is between +15 percent and +25 percent is calculated like this:

$$P(0.15 \le X \le 0.25) = P\left(\left(\frac{0.15 - 0.05}{0.10}\right) \le Z \le \left(\frac{0.25 - 0.05}{0.10}\right)\right)$$
$$= P(1.00 \le Z \le 2.00)$$
$$= P(Z \le 2.00) - P(Z \le 1.00)$$
$$= 0.9772 - 0.8413$$
$$= 0.1359$$

This shows that if the returns to the portfolio are normally distributed with a mean of 5 percent and a standard deviation of 10 percent, then 13.59 percent of the returns should be between +15 percent and +25 percent.

Based on these probabilities, the *expected frequencies* of the portfolio returns within each category equal the probability that a return falls into the category times the size of the portfolio. Expected frequency for returns between −15 percent and −5 percent is therefore

$$-15\% \text{ to } -5\%: \ 0.1359 \times 100 = 13.59$$

Because the probability that a return is between −15 percent and −5 percent, in a portfolio of 100 stocks the return should be in this range 13.59 times.

The rest are computed in a similar manner:

$$-5\% \text{ to } +5\%: \ 0.3413 \times 100 = 34.13$$
$$+5\% \text{ to } +15\%: \ 0.3413 \times 100 = 34.31$$
$$+15\% \text{ to } +25\%: \ 0.1359 \times 100 = 13.59$$

Table 8-2 shows the observed frequencies and the expected frequencies for the portfolio.

Table 8-2   Observed and Expected Frequency of Portfolio Returns

Returns	*−15% to −5%*	*−5% to +5%*	*+5% to +15%*	*+15% to +25%*
Observed frequency	22	29	37	12
Expected frequency	13.59	34.13	34.13	13.59

Even though it's impossible for a fractional number of returns to fit within each of these categories, the expected frequencies are *not* rounded with a goodness of fit test.

Based on these results, the *test statistic* for the goodness of fit test is constructed as follows:

$$\chi^2 = \frac{\sum_{j=1}^{k}(O_j - E_j)^2}{E_j}$$

In this formula:

O_j = Observed frequency in category j

E_j = Expected frequency in category j

k = Number of categories

χ^2 is used to indicate that this test statistic follows the *chi-square* distribution.

χ is the Greek letter *chi* (pronounced "ki").

The formula for the test statistic shows that the closer the observed and expected frequencies are, the less likely the null hypothesis is to be rejected.

In this example, the test statistic is computed as follows:

$$\chi^2 = \frac{(22-13.59)^2}{13.59} + \frac{(29-34.13)^2}{34.13} + \frac{(37-34.13)^2}{34.13} + \frac{(12-13.59)^2}{13.59} = 6.402$$

The critical value

The *critical value* for the goodness of fit test is taken from the chi-square distribution, with df (degrees of freedom) equal to $k-1-m$. This is expressed as follows:

$$\chi^2_{k-1-m,\alpha}$$

In this expression, α represents the level of significance.

There is only a single critical value, because a goodness of fit test is always implemented as a right-tailed test.

For this critical value, the df is computed as $k-1-m$.

- k represents the number of categories.
- m refers to the number of values that must be estimated from the sample data.

In this case, $k = 4$ because there are four categories of returns. m represents the number of *parameters* of the assumed probability distribution that must be estimated from sample data. For the normal distribution, there are *two* parameters: the mean (μ) and the standard deviation (σ). If you don't specify the values of these parameters in the null hypothesis, then you must estimate them by computing the sample mean and the sample standard deviation.

In this case, the mean of 5 percent and standard deviation of 10 percent are both specified in the null hypothesis. Therefore, the sample mean and sample standard deviation don't have to be computed. Because neither parameter must be estimated from sample data, $m = 0$. The degrees of freedom equals the following:

$$k - 1 - m = 4 - 1 - 0 = 3$$

You can find the critical value in a chi-square table. Table 8-3 shows a portion of the chi-square table.

Table 8-3			The Chi-Square Distribution				
df\Right-tail Area	0.99	0.975	0.95	0.90	0.10	0.05	0.025
1	0.000	0.000	0.000	0.016	2.706	3.841	5.024
2	0.020	0.051	0.103	0.211	4.605	5.991	7.378
3	0.115	0.216	0.352	0.584	6.251	7.815	9.348
4	0.297	0.484	0.711	1.064	7.779	9.488	11.143
5	0.554	0.831	1.145	1.610	9.236	11.070	12.833
6	0.872	1.237	1.635	2.204	10.645	12.592	14.449
7	1.239	1.690	2.167	2.833	12.017	14.067	16.013
8	1.646	2.180	2.733	3.490	13.362	15.507	17.535

The top row of the table represents areas in the *right* tail of the chi-square distribution. The first column represents the number of degrees of freedom (df).

In this case, the right-tail area is 0.05, with 3 df. This is because the level of significance is 0.05 (5 percent), and the df equals $k - 1 - m$, or 3.

By looking in the row corresponding to 3 df and the column corresponding to a right-tail area of 0.05, you will find the appropriate critical value of 7.815.

The decision

Because this is a right-tailed test, in order to reject the null hypothesis, the test statistic must *exceed* the critical value. For this type of test, rejecting the null hypothesis would indicate that the population is *not* normally distributed.

The test statistic is 6.402, whereas the critical value is 7.815. Therefore, the null hypothesis is not rejected. This indicates that it is reasonable to assume the population is normally distributed with the specified parameters: mean = 5 percent and standard deviation = 10 percent.

If the null hypothesis had been rejected, it could have been for one of several reasons:

- ✔ The population is normal, and the mean is 5 percent, but the standard deviation isn't 10 percent.

- ✔ The population is normal, and the standard deviation *is* 10 percent but the mean isn't 5 percent.

- ✔ The population is normal, but the mean isn't 5 percent and the standard deviation isn't 10 percent.

- ✔ The population isn't normal.

There is a lot of information packed into the null hypothesis. It's a fairly strong claim about the distribution. But as you can see, if the goodness of fit test leads you to reject it, it doesn't give you a very clear sense of why.

Jarque-Bera test

Another test that you can use to determine whether a population is normally distributed is the Jarque-Bera test. The *Jarque-Bera test* is based on a comparison of the skewness and kurtosis of a population to those of the normal distribution.

Skewness

Skewness measures the degree of *asymmetry* of a probability distribution. (See Chapter 5 for a fuller discussion of skewness.) For the normal distribution, skewness equals zero, since the normal distribution is *symmetrical* about its mean.

For a symmetrical distribution, the area above the mean is a *mirror image* of the area below the mean.

Kurtosis

Kurtosis measures the likelihood of extreme outcomes in a distribution relative to the normal distribution. (See Chapter 5 for a more detailed discussion of kurtosis.) For the normal distribution, kurtosis equals 3. A distribution with a kurtosis exceeding 3 is said to be a *fat-tailed* distribution, and a distribution whose kurtosis is less than 3 is said to be a *skinny-tailed* distribution.

Excess kurtosis

A related measure is known as *excess kurtosis*. This is defined as follows:

excess kurtosis = kurtosis − 3

Table 8-4 shows how distributions can be classified with the excess kurtosis measure.

Table 8-4	Excess Kurtosis
Excess kurtosis > 0	Fat-tailed distribution
Excess kurtosis = 0	Normal distribution
Excess kurtosis < 0	Skinny-tailed distribution

The Jarque-Bera test uses the skewness and kurtosis of a sample to create a test statistic that can be used to determine whether your data is approximately normally distributed.

The null and alternative hypotheses

In the case of the Jarque-Bera test, the null and alternative hypotheses would be the following:

H_0: The population is normally distributed.

H_1: The population is not normally distributed.

As with the goodness of fit test, this is always a right-tailed test.

Computing the test statistic

For the Jarque-Bera test, the test statistic is a function of the skewness and kurtosis of a sample drawn from the underlying population. The test statistic is computed as follows:

$$JB = \frac{n}{6}\left(S^2 + \frac{1}{4}(K-3)^2\right)$$

In this formula:

JB = Jarque-Bera test statistic

n = Sample size

S = Sample skewness

K = Sample kurtosis

Because K represents sample kurtosis, $K - 3$ represents sample *excess* kurtosis.

For example, suppose a sample of 24 is drawn from a population of returns to a stock over the past 20 years. The sample skewness is determined to be –1, and the sample kurtosis equals 4. The Jarque-Bera test statistic is as follows:

$$JB = \frac{n}{6}\left(S^2 + \frac{1}{4}(K-3)^2\right) = \frac{24}{6}\left((-1)^2 + \frac{1}{4}(4-3)^2\right) = 5$$

The critical value

The Jarque-Bera test statistic follows the chi-square distribution with 2 degrees of freedom. The test statistic can be written like this:

$$\chi_2^2$$

In this example, assume that the level of significance is 0.05. Therefore, you find the critical value in Table 8-3 to be the following:

$$\chi_2^2 = 5.991$$

Because this is a right-tailed test, in order to reject the null hypothesis, the test statistic must *exceed* the critical value. As with the goodness of fit test, rejecting the null hypothesis would indicate that the population is *not* normally distributed.

The test statistic is 5, and the critical value is 5.991. Therefore, the null hypothesis is *not* rejected. This indicates that the population may be reasonably assumed to be normally distributed.

It should be noted that in the case of fairly small samples (generally anything under 30 is considered small), the Jarque-Bera test should be avoided. The test statistic does not really start to look like a true χ^2_2 distribution until the sample size gets larger.

As you make your way through the remainder of this book, you will come across a variety of approaches to making statistically based predictions. Frequently, we point out that these approaches are partly based on the assumption that the variables you're using are approximately normally distributed. This chapter has given you some tools to verify (or not) the assumption of normality.

Chapter 9

Dealing with Missing or Incomplete Data

In This Chapter

▶ Seeing the different ways in which observations can be missing from a dataset

▶ Understanding what types of problems can be caused by missing data

▶ Learning how to overcome the problems caused by missing data

M issing data is a major problem in all areas of statistical analysis. Incomplete information can make it impossible to use many types of statistical techniques; for example, a paired t-test cannot be run unless there are equal numbers of observations for two variables.

You use a paired t-test to test the hypothesis that the means of two different populations are equal to each other.

Missing observations can severely distort the results of any statistical procedure, calling into question the validity of the results. Fortunately, many new techniques have been developed in recent years to manage the problem of missing data. The use of these techniques has accelerated, partially due to the development of highly sophisticated statistical software packages.

This chapter introduces several potential causes for missing data and explains several techniques for handling the problem. Solutions range from the very simple, such as averaging techniques, to the very sophisticated, including simulation techniques and maximum likelihood estimation.

Missing Data: What's the Problem?

The term *missing data* refers to a situation where some observations have not been included in a dataset. For example, suppose that a time series of Apple stock prices for the year 2014 doesn't show a price for Monday, October 27,

2014. This would be considered a missing value since the stock market was open on that date and Apple stock was traded heavily during the day.

There are several potential causes of missing observations, which can be classified as nonresponse and missingness.

Missingness doesn't appear in most dictionaries, but it does appear on the Wiktionary website (`http://en.wiktionary.org/wiki/missingness`). It's one of those technical words that will sooner or later be absorbed by the English language.

Nonresponse

Nonresponse refers to a situation where an observation is missing from a dataset because the information was intentionally not reported. For example, if data is obtained from a survey, nonresponse could be caused by the unwillingness of participants to provide certain types of information. This is a significant problem for survey data, because the accuracy of the results depends on having a complete set of valid responses.

Nonresponse can lead to the problem of *non-response bias*. This is a situation in which the statistical results of a study can be distorted if there is a systematic pattern to the missing observations. For example, if a survey is conducted to determine the average income in a given geographic area, and many of the high-income households do not bother to respond, the results will be biased downward — that is, the estimated average income will be lower than the actual average income.

Missingness

Missingness refers to cases where observations have been left out of a dataset for one of several possible reasons; this could be due to random chance, technical errors, human errors, and so on. For example, a stock price could be missing from a dataset due to computer error, a bad data feed from the stock exchange, and so forth.

There are a few different types of missingness:

- ✔ Missing completely at random (MCAR)
- ✔ Missing at random (MAR)
- ✔ Missing not at random (MNAR)

Missing completely at random (MCAR)

Data is said to be missing completely at random (MCAR) when there's no particular pattern to the missing observations. For example, if a small number of Apple stock prices are missing from a time series of 2014 prices, and there's no pattern to the missing dates or prices, then the data is considered to be MCAR. On the other hand, if the missing data always occurs on a Friday, then there's a pattern to the missing data that must be investigated; the data is clearly not MCAR.

For the Apple stock example, it's assumed that the prices are a function of time; the prices are considered to be the dependent variable *(Y)* and time is the independent variable *(X)* in this dataset. Because the missingness of *Y* is not related to the values of *X* or *Y,* this is a case of data that is MCAR.

With data that is missing completely at random, the probability of being missing is the same for all observations. Of the three types of missingness, MCAR is the simplest to correct for.

Missing at random (MAR)

When data are *missing at random,* there is no relationship between the missingness of *Y* and the values of *Y.* Using the example of the Apple stock prices, suppose the missing values are more likely to occur early in the week than later in the week, but are unrelated to the stock prices themselves. Because the missingness is related to time *(X)* but not to price *(Y),* the data is considered to be MAR.

Missing not at random (MNAR)

When there is a pattern to the missing data, it is said to be *missing not at random.* In other words, the missingness of *Y* is a function of the value of *Y* and/or *X.* For example, if the price of Apple stock is missing on the first day of every month, or the missing values are the largest prices in the time series, or some combination of the two, the data is considered to be MNAR.

One of the major drawbacks of MNAR data is that in order to run traditional statistical techniques with the data, an explicit model of the missingness of the data is required to fill in the gaps. This can be an extremely complex process and can introduce new errors into statistical analysis.

One model that has been successfully used for regression analysis with missing not at random data is the Heckman sample selection model. (For his contributions to econometrics, James Heckman was awarded the Nobel Prize in Economics in 2000.)

Techniques for Dealing with Missing Data

You can use several methods to deal with missing data. In some cases, you remove missing data from a dataset; in other cases, you use a specialized technique to provide estimates for the missing values. Another technique that you may use is known as *maximum likelihood estimation*. With this technique, you estimate a function that identifies the most likely properties of the missing data.

Deletion techniques

Here are two methods that involve the removal of incomplete data:

- Casewise deletion
- Pairwise deletion

Casewise deletion

With *casewise deletion* (also known as *complete-case analysis*), incomplete data is deleted from the dataset. For example, suppose a multivariate equation is estimated with regression analysis. The dependent variable *(Y)* is annual salary, and the independent variables are years of experience (X_1) and years of graduate education (X_2). The data is gathered by randomly selecting employees at a large corporation and recording their salary, experience, and graduate education. If any combination of salary, years of experience, and years of graduate education is missing for a given employee, that employee's information is deleted from the dataset. Any statistical analysis that is performed is based only on complete data.

Table 9-1 shows a sample of results for this corporation.

Table 9-1	Employee Salaries, Experience, and Education		
Employee Number	**Annual Salary (Y) ($ Thousands per Year)**	**Years of Experience (X1)**	**Years of Graduate Education (X2)**
1	100	12	3
2	95	8	***
3	122	14	2
4	88	***	3
5	***	15	5
6	72	4	2

The table shows that one variable is missing for employees 2, 4, and 5. Using the casewise deletion method, you delete all information about these employees from the dataset.

One of the main drawbacks to this approach is that it can significantly reduce the size of the sample being used, which can be problematic if the sample size is already small. One of the advantages of this approach is that if the data is MCAR, then this approach doesn't introduce any bias into any statistical analysis that is performed on the data.

Bias results when the expected or average value of an estimate is consistently above or below the value being estimated. For example, a procedure that's designed to estimate the mean age of the residents of a state that persistently overestimates the mean age is said to be *upward biased*. A procedure that persistently underestimates the mean age is said to be *downward biased*. A procedure that is correct on average is said to be *unbiased*.

Pairwise deletion

With *pairwise deletion* (also known as available-case analysis), all existing information in a dataset is used to compute correlations that are required to estimate the coefficients of a regression equation.

For example, suppose that you use a multiple regression equation to measure the relationship between a major league baseball team's wins during a season (Y) and runs scored (X_1), earned runs allowed (X_2) and saves (X_3). A sample of this data is shown in Table 9-2.

Table 9-2	Baseball Runs, Earned Runs, and Saves			
Major League Team	Wins (Y)	Runs Scored (X1)	Earned Runs Allowed (X2)	Saves (X3)
1	84	820	745	39
2	90	880	730	44
3	86	830	750	***
4	68	710	***	28
5	91	858	706	45
6	***	755	890	40
7	100	905	685	51
8	95	872	722	48

The value of Y is missing for team 6; because the data for X_1, X_2, and X_3 is present, the team's data may be included in the calculation of the correlations between X_1 and X_2, X_1 and X_3, and X_2 and X_3. Similarly, the value of X_2 is missing for team 4. Since the data for Y, X_1, and X_3 is present, the team's data may be used to compute the correlations between Y and X_1, Y and X_3, and X_1 and X_3.

In addition to correlations, all available data is used to compute the means, variances, and standard deviations of the dependent and independent variables. With this information, you can perform many different statistical techniques on the data, including regression analysis.

One of the advantages of pairwise deletion is that it uses all available information, so it doesn't reduce the sample size. The main drawback is that the correlations, means, variances, and standard deviations are all being computed from samples of different sizes, so that their values may not be consistent with each other.

Due to the large number of potential problems with pairwise deletion, it's sometimes referred to by statisticians as "*un*wise deletion."

In general, if the amount of missing data is very small and is representative of the overall dataset, deletion techniques can work very well. Otherwise, you should use another technique, such as imputation, discussed next.

Imputation techniques

Statistical methods that produce estimates for missing data are known as *imputation* techniques. In these cases, missing values are estimated or "imputed" from the rest of the dataset. Imputation techniques can be divided into simple imputation techniques and multiple imputation techniques.

Simple imputation techniques

With a *simple imputation* technique, a missing value is estimated from the remaining values in the dataset. There are several different types of simple imputation techniques, including the following:

- Mean substitution
- Regression substitution
- Hotdecking

Mean substitution

With *mean substitution,* you can estimate a missing value as the mean of the non-missing values of the variable. For example, Table 9-3 shows a sample of six observations taken on variables X and Y.

Table 9-3	Estimating a Value with Mean Substitution	
Observation	*Y*	*X*
1	10	8
2	14	9
3	12	7
4	10	6
5	8	5
6	***	7

The value of Y is missing for observation number 6. Rather than deleting this observation, the value of Y is estimated as the mean of the other values of Y:

$$\frac{10+14+12+10+8}{5} = 10.8$$

Mean substitution is not generally a very good solution to missing data, particularly when there are a lot of missing values. The problem with taking averages is that it eliminates any differentiation between the observations that have the missing values. And differentiation among observations is the bread and butter of predictive statistics. The techniques described in this section are much friendlier to statistical prediction.

In some circumstances, you can replace missing values with other observations in a dataset. For example, in a time series of stock prices, suppose that there's no price for a stock on Tuesday, January 14. If the stock didn't trade during the day, you can use the price for January 13 as a proxy for the January 14 price.

Another option is to use *interpolation.* This is a technique in which a single observation in a dataset is estimated from other observations. For example, if the price of a stock is $100 on January 13 and $105 on January 15, you can interpolate value for January 14 simply by averaging the prices on January 13 and January 15:

$$\left(\frac{100+105}{2}\right) = 102.5$$

Regression substitution

With *regression substitution,* you can estimate a missing value by running a regression on the remaining data. For example, based on Table 9-3, a regression is run using the first five observations. The estimated regression equation is

$$\hat{Y}_i = 2.4 + 1.2X_i$$

Since $X = 7$ for the sixth observation, the predicted value of Y is

$$\hat{Y}_i = 2.4 + 1.2X_i = 2.4 + 1.2(7) = 10.8$$

Therefore, you use 10.8 as the estimated value of Y for the sixth observation in the dataset.

Regression substitution is a useful technique for data that is either MCAR or MAR.

Hotdecking

With the *hotdecking* approach, you can estimate a missing value by comparing the values of the remaining variables with the rest of the dataset and finding a match. For example, in Table 9-3, the sixth observation is missing a value for Y and has a value of 7 for X. The third observation also has a value of 7 for X, along with value of 12 for Y. Using the hotdecking approach, the estimated value of Y for the sixth observation would therefore be 12.

One of the drawbacks of hotdecking is the possibility that there are no matches in the data. Another potential problem arises if there are multiple missing values. This technique is also based on the assumption that the data is either MCAR or MAR.

Multiple imputation

Multiple imputation is a technique in which Monte Carlo simulation is used to generate estimated values for missing data. Multiple imputation is implemented as a series of iterations. Within each iteration, estimated values are imputed for missing data using simulation techniques and then statistical analysis is performed on the data. The process is then repeated, with new values being simulated for the missing data. The same statistical analysis is performed again. After each round, the estimates of the missing data should converge to the correct values. After a series of iterations, the results of the statistical analysis during each iteration are combined to provide a single result.

One of the drawbacks to multiple imputation is that the simulation technique being used depends on the assumed statistical properties of the missing data. In many cases, the data are assumed to be normally distributed. If this isn't accurate, then the reliability of the multiple imputation process becomes problematic.

Expectation-maximization (EM)

Expectation-maximization (EM) is an alternative to deletion and imputation techniques. You implement EM in two steps:

- ✔ Estimation
- ✔ Maximization

With the estimation step, you estimate several summary measures (such as means, variances, covariances, and other statistical measures) with regression analysis. This is followed by the maximization step. Maximum likelihood estimation (MLE) is used to provide estimated values for the missing data.

Once the missing data has been estimated, the EM process returns to the estimation step. Here, you refine the original estimates of the means, variances, covariances, and so forth by re-estimating them with regression analysis, including the estimated values for the missing data. This is followed by a repeat of the maximization step. At this step, the estimates for the missing data are updated with maximum likelihood estimation. The process continues until the results have reached their final values.

One drawback to the expectation-maximization technique is that it can give biased results. An advantage is that it can be used as the starting point for more advanced techniques.

Chapter 10

Sending Out a Posse: Searching for Outliers

In This Chapter

▶ Learning how to identify outliers with formal statistical procedures

▶ Seeing how outliers affect statistical tests

▶ Finding out how to avoid the problems associated with outliers

An *outlier* is a member of a dataset that's significantly larger or smaller than the other values in the dataset. Outliers can appear in all walks of life. For example, the following would be considered outliers:

▶ A man who is seven feet tall

▶ A woman who is 100 years old

▶ A household that has an annual income of $100 million per year

▶ A baseball player who hits .400 during an entire season

In statistical analysis, an outlier refers to a value that is substantially different from the other values within a sample or a population. For example, suppose you take a sample of housing prices in a small town, with the following results (in hundreds of thousands of dollars):

240, 270, 290, 305, 332, 348, 371, 404, 2,250

In this case, you would consider the home that's worth $2.25 million to be an outlier because it's so much more expensive than the other homes in that town. In fact, it's more than five times as costly as the next most expensive home.

Outliers are a cause for concern because they may indicate a problem with the data being used. For example, an outlier could result from an error in gathering the data, which calls into question the accuracy of the entire dataset. In the housing prices example, the 2,250 figure could be a typo; it may be that the correct figure is 225, which is much more in line with the remaining prices. The presence of outliers can also affect the choice of statistical procedure that you use to analyze the data. Outliers can greatly affect the accuracy of some statistics, so it's important to choose the right procedure.

In the case of our house price example, even the simple calculation of the mean can be greatly affected by an outlier. If the 2 million-dollar house is left in the calculation, the mean turns out to be just over $534,000. This is actually significantly higher than eight of the nine observations. So it really isn't indicative of any central tendency. If the large observation is assumed to be a typo and replaced with $225,000, the mean drops all the way to $309,000.

A statistical procedure that isn't greatly affected by outliers is said to be *robust*. As the previous example illustrates, the mean is not a robust statistic. There are two other measures of central tendency from Chapter 5 that are robust: The mode and the median are not affected by outliers.

This chapter shows several formal procedures for identifying outliers and discusses how to overcome the problems that may arise in their presence.

Testing for Outliers

Identifying outliers isn't a cut-and-dried matter. There can be disagreement about what does and does not qualify as an outlier. For example, some researchers may define an outlier as any observation that is three or more standard deviations away from the mean (the notion of standard deviation is discussed in detail in Chapter 5), whereas others may use a definition of four or more standard deviations away from the mean.

To properly identify outliers, you should use more formal techniques, such as the following:

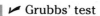 Grubbs' test

✔ Chi-square test

✔ Dixon's Q test

Prior to using these formal tests, the first step in identifying outliers is to visually inspect the data with a *graph* to see if there are any potential candidates for outliers in the data.

Graphical tests of outliers

The definition of an outlier depends on the assumed probability distribution of a population. For example, as discussed at length in Chapter 8, you assume that the values of a population are normally distributed. If population really is normally distributed, the graph of a dataset should have the same signature bell shape — if it doesn't, that could be a sign that there are outliers in the data.

You may use three graphical techniques to identify outliers:

- Histograms
- Box plots
- QQ-plots

Histograms

A *histogram* is a graph used to visually represent a probability distribution with a series of vertical bars. The horizontal axis shows values or ranges of values for the variable being studied, and the vertical axis shows the corresponding frequencies of these values.

As an example, the Standard and Poor's 500 index (S&P 500) is a stock market index that represents the prices of the 500 largest U.S. stocks, weighted by their market capitalization. A stock's *market capitalization* equals the price per share times the number of shares outstanding.

Figure 10-1 shows a histogram of the daily returns for the Standard and Poor's 500 stock market index during the years 2009–2013.

Figure 10-1: A histogram of the daily returns to the S&P 500 from 2009–2013.

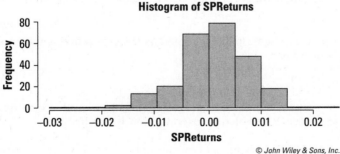

© John Wiley & Sons, Inc.

According to this histogram, most of the returns were close to zero during this period. Returns above 0.01 (1 percent) or below –0.01 (–1 percent) occurred relatively infrequently. However, for the returns that did occur outside the small range around 0, the occurrence of negative returns outweighed the occurrence of positive returns, as seen by the extreme length of the left tail.

The shape of the histogram shows that the distribution of returns to the Standard and Poor's 500 during this period is unlikely to be normal. One problem is that the normal distribution is symmetrical about its mean, whereas the histogram shows that the distribution of returns is *negatively skewed* (that is, there's an imbalance between negative and positive returns, with more negative than positive returns).

Box plots

A *box plot* shows the distribution of a dataset in a box. The box is based on *quartiles,* which are like percentiles except that there are only four of them. The box plot is structured as follows:

- ✔ The top of the box represents the *third quartile* (or upper quartile) (Q_3) of the data. This is equivalent to the 75th percentile.

- ✔ The bottom of the box represents the *first quartile* (or lower quartile) (Q_1) of the data. This is equivalent to the 25th percentile.

- ✔ The middle of the box (shown with a line) represents the *second quartile* (Q_2) of the data (also known as the *median*).

The first quartile of a dataset is a value that is greater than 25 percent of the elements of the dataset and less than the remaining 75 percent. The second quartile (that is, the median) is a value that is greater than 50 percent of the elements and less than the remaining 50 percent. The third quartile is a value that is greater than 75 percent of the elements and less than the remaining 25 percent.

The *interquartile range* (IQR) is defined as the difference between the third and first quartiles:

$$IQR = Q_3 - Q_1$$

The *IQR* is used as a measure of *dispersion,* or how spread out the data is about the center. It can also be used to identify outliers.

For a box plot, there are lines above and below the box. The top line represents the maximum value in a dataset, excluding outliers. The bottom line represents the minimum value in a dataset, again excluding outliers. The individual points shown above and below these lines are the outliers in the dataset.

When you're using a box plot, an outlier is defined as follows:

✔ If a data point is below $Q_1 - 1.5(IQR)$, it is considered to be an outlier.

✔ If a data point is above $Q_3 + 1.5(IQR)$, it is considered to be an outlier.

Figure 10-2 shows a box plot of the daily returns to the S&P 500 stock market index during the years 2009–2013.

Figure 10-2:
A box plot of the daily returns to the S&P 500 from 2009–2013.

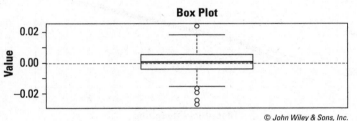

© John Wiley & Sons, Inc.

The box plot shows that there is one outlier that is significantly greater than the rest of the returns in the dataset. There are also four outliers that are significantly smaller than the rest of the returns in the dataset. The existence of these outliers shows that the dataset may not be normally distributed.

QQ-plots

You can plot sample data with a *QQ-plot* (short for quantile-quantile plot). This plot compares the quantiles of the sample data with the quantiles of a specified probability distribution, such as the normal.

Quantiles are used to divide a dataset into equally sized groups based on the value of a particular numeric variable. There are several types of quantiles, including the following:

✔ **Percentiles** divide a dataset into 100 equal groups, each corresponding to a percentage of the total. For example, if a group of 1,000 students takes a standardized exam, and 200 of them receive a score below 300, then 300 would be the 20th percentile of this dataset. This indicates that 20 percent of students scored below 300, whereas the remaining 80 percent scored higher than 300.

✔ **Deciles** divide a dataset into ten equal groups, each representing 10 percent of the total. For example, the 4th decile corresponds to the 40th percentile.

✔ **Quartiles** divide a dataset into four equal groups, each representing 25 percent of the total. For example, the 3rd quartile corresponds to the 75th percentile.

Figure 10-3 shows a QQ-plot of the daily returns to the S&P 500 stock market index during 2009–2013, compared with the normal distribution:

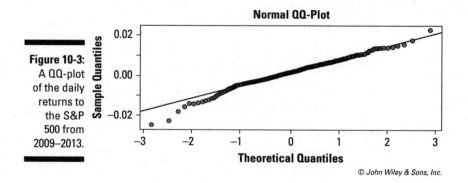

Figure 10-3: A QQ-plot of the daily returns to the S&P 500 from 2009–2013.

© John Wiley & Sons, Inc.

The solid line on the graph represents the quantiles of the normal distribution. 0 represents the mean; therefore, half of the values are below 0, and half are above it. About 95 percent of the values are below 2 (2 represents two standard deviations above the mean), whereas 5 percent of the values are below –2 (–2 represents two standard deviations below the mean). If the S&P returns were normally distributed, their quantiles should lie on the line.

The points on the graph are the actual observations in the S&P 500 dataset. For the normal quantiles that are greater than 2 (that is, two standard deviations above the mean), the S&P 500 returns are above the line, which indicates that the right tail is too "fat" to be consistent with the normal distribution. For normal quantiles that are below –1 (that is, one standard deviation below the mean), the S&P 500 returns are below the line, which indicates that the left tail is also too fat to be consistent with the normal distribution.

Overall, the distribution of returns to the S&P 500 appears to be a fat-tailed distribution, meaning that extreme outcomes are much more likely than would be the case with the normal distribution.

Hypothesis tests for outliers

In additional to graphical tests, there are several formal statistical tests that are designed to detect outliers. Three of these take the form of hypothesis tests.

You may recall from Chapter 5 that a hypothesis test is a procedure for determining whether a proposition can be rejected based on sample data. Hypothesis tests always involve comparing a test statistic from the data to an appropriate distribution to determine whether a given hypothesis is supported by the data. The following sections give a variety of hypothesis tests related to the detection of outliers.

Grubbs' test

With a Grubbs' test, you assume that the dataset being tested for outliers is normally distributed. (Chapter 8 is dedicated to the subject of how to go about verifying this assumption.) The null and alternative hypotheses are as follows:

H_0: There are no outliers.

H_1: There is at least one outlier.

The test statistic is as follows:

$$G = \frac{\max |Y_i - \bar{Y}|}{s}$$

where

G = The test statistic for the Grubbs' test

Yi = A single element in the dataset being tested

Y = The sample mean

s = The sample standard deviation

The test statistic produces the sample element that is furthest from the sample mean (positive or negative) expressed as standard deviations. For example, if the sample mean is 5, the largest sample element is 11, and the sample standard deviation is 2, then the test statistic would be $(11 - 5)/2 = 6/2 = 3$ standard deviations away from the mean.

The critical value is as follows:

$$\frac{n-1}{\sqrt{n}}\sqrt{\frac{t^2}{n-2+t^2}}$$

Where

n is the size of the sample drawn from the population.

t is a value drawn from the Student's t-distribution; it has a right tail area equal to the level of significance (α) and $n-2$ degrees of freedom (df).

The test can be conducted to determine whether there is an outlier, whether the maximum value is an outlier, whether the minimum value is an outlier, and so on.

For example, the following shows the results of applying Grubbs' test to the S&P 500 returns from 2009–2013. The test is conducted to find a single outlier. Grubbs' test results for one outlier:

Data: SPReturns

G = 3.8509, U = 0.9404, p-value = 0.01177

Alternative hypothesis: Lowest value –0.0253283545257448 is an outlier

With a level of significance equal to 0.05, and a p-value of 0.01177, the p-value is below the level of significance. Therefore, the null hypothesis of no outliers is rejected. Furthermore, the test indicates that the minimum value in the dataset is an outlier.

Chi-square test

You can test for outliers with the chi-square distribution. The null and alternative hypotheses are as follows:

H_0: There are no outliers.

H_1: There is at least one outlier.

The test statistic is based on the differences between the actual members of a dataset and the corresponding members of an assumed probability distribution, such as the normal.

For example, the following shows the results of applying the chi-square test to the S&P 500 returns from 2009–2013:

Chi-square test for outlier

Data: SPReturns

X-squared = 14.8292, p-value = 0.01177

Alternative hypothesis: Lowest value –0.0253283545257448 is an outlier

With a level of significance equal to 0.05, and a p-value of 0.01177, the p-value is below the level of significance. Therefore, the null hypothesis of no outliers is rejected. Furthermore, the test indicates that the minimum value in the dataset is an outlier.

Dixon's Q test

With Dixon's Q test, you assume the dataset being tested for outliers is normally distributed. The null and alternative hypotheses are as follows:

H_0: There are no outliers.

H_1: There is at least one outlier.

The test statistic is as follows:

$$Q = \frac{gap}{range}$$

Gap refers to the absolute value of the difference between an outlier and the next closest value in the dataset. *Range* refers to the difference between the largest value in the dataset and the smallest value in the dataset.

One of the drawbacks to Dixon's Q test is that you can apply it only to a sample containing between 3 and 30 observations.

The following shows the results of applying Dixon's Q test to the S&P 500 returns during the first 30 trading days of 2009:

Dixon test for outliers

Data: SPR

Q = 0.4359, p-value = 0.03185

Alternative hypothesis: Lowest value –0.0116057775514049 is an outlier

With a level of significance equal to 0.05, and a p-value of 0.03185, the p-value is below the level of significance. Therefore, the null hypothesis of no outliers is rejected. Furthermore, the test indicates that the minimum value in the dataset is an outlier.

Robust Statistics

As mentioned, a statistic is said to be *robust* if it isn't strongly influenced by the presence of outliers. For example, the mean is not robust because it can be strongly affected by the presence of outliers. On the other hand, the median *is* robust — it isn't affected by outliers.

For example, suppose the following data represents a sample of household incomes in a small town (measured in thousands of dollars per year):

32, 47, 20, 25, 56

You compute the sample mean as the sum of the five observations divided by five:

$$\frac{32 + 47 + 20 + 25 + 56}{5} = \frac{180}{5} = 36$$

The sample mean is $36,000 per year. Most of the households in the sample are very close to this value.

Suppose instead that the sample consists of the following values:

32, 47, 20, 25, 376

Because the household income of $376,000 is substantially greater than the next closest household income of $32,000, the household income of $376,000 can be considered to be an outlier.

With the outlier, the sample mean is now as follows:

$$\frac{32 + 47 + 20 + 25 + 376}{5} = \frac{500}{5} = 100$$

This measure isn't representative of most of the households in the town. Thus the usefulness of the mean is compromised in the presence of outliers.

You compute the median of the sample by sorting the data from lowest to highest and then finding the value which divides the sample in half. In other words, half of the observations are below the median, and half are above.

The first sample:

32, 47, 20, 25, 56

The sorted sample:

20, 25, 32, 47, 56

In this case, the median is 32 because half of the remaining observations are below 32 and half are above it.

The second sample:

32, 47, 20, 25, 376

The sorted sample:

20, 25, 32, 47, 376

Despite the presence of the outlier of 376, the median is still 32. It hasn't been affected by the outlier. This shows that unlike the mean, the median is *robust* with respect to outliers.

Other examples of robust statistics include the median, absolute deviation, and the interquartile range.

Dealing with Outliers

Once you have identified outliers, the question becomes what to do about them. There is no one answer to this question. And your strategy for dealing with outliers depends on your familiarity with whatever field you are working in.

It may become clear to you that an outlier is simply a data entry or other type of error. If so, you can try to correct the problem. If there is no obvious correction to be made, then it's probably best to just delete the observation.

Your outlier may, however, reflect legitimate — that is, correct — data. In that case, you probably shouldn't be so quick to toss it out. A better strategy is to perform your analysis twice, once with and once without the outlier included. Doing that allows you to evaluate how much the outlier is affecting your results. If you're building a predictive model, it enables you to evaluate the effectiveness of the two models side by side.

In any case, looking for outliers should be standard operating procedure when you're beginning your analysis of a dataset. Like the other data preparation procedures discussed in Part II of this book, recognizing and dealing with outliers will positively affect your results.

And now we have (finally!) come to the end of the preliminaries. In the next two parts of this book, we get to the main event at last. We begin to discuss how to analyze data in a way that generates insight and predictions about what is going on in the world.

Part III
Exploratory Data Analysis (EDA)

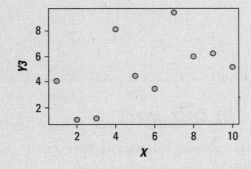

In this part . . .

- Exploring Exploratory Data Analysis (EDA)
- Getting visual with graphical techniques
- Understanding univariate and multivariate statistical techniques
- Going further with regression analysis and time series analysis

Chapter 11

An Overview of Exploratory Data Analysis (EDA)

In This Chapter

▶ Seeing how the focus of EDA is different from traditional statistical analysis

▶ Exploring important graphical EDA techniques

▶ Understanding key quantitative EDA techniques

*E*xploratory data analysis (EDA) is an approach to data analysis that lets you determine a dataset's properties so you can use the appropriate technique or techniques for analyzing the data. This helps ensure that you won't impose assumptions on the data that aren't warranted. Unlike more traditional approaches, which impose a specific model on a dataset based on predetermined assumptions, with EDA the structure of a dataset determines which techniques should be used to analyze the data.

The field of EDA was introduced in 1977 by the mathematician John Tukey. Tukey believed that it was important to let a dataset determine the types of analysis that should be performed on it, rather than look for confirmation of predetermined assumptions.

EDA is designed to accomplish several important objectives:

✔ Understanding the properties of a dataset

✔ Determining what type of structure, if any, is present in a dataset

✔ Identifying outliers

✔ Determining which probability distribution the data follows

EDA is used to determine several key properties of a dataset. These include *summary measures* such as these:

- Mean
- Median
- Mode
- Variance
- Standard deviation

These measures provide a great deal of information about the data in a very compact form. The mean, median, and mode are measures of central tendency and describe the location of the data. The variance and standard deviation are measures of dispersion and describe how "spread out" the data is relative to the center.

EDA can also be used to determine whether there is any structure in the data — that is, to determine if there are any particular patterns in the data. With EDA, it's relatively easy to spot potential *outliers,* which are values that are far removed from other values in the dataset. EDA can also help determine whether a dataset follows a particular probability distribution, such as the binomial, normal, Student's t-distribution, and so on.

Visualizing your data is often enlightening. Insights can be gained at a glance from various graphs of data. And these graphs can raise questions that show you where you need to dig more deeply into your data.

This chapter explores two basic types of techniques for EDA:

- Graphical techniques
- Quantitative techniques

Graphical EDA Techniques

EDA is based heavily on graphical techniques. (Graphical techniques are discussed extensively in Chapter 12.) You can use graphical techniques to identify the most important properties of a dataset.

Here are some of the more widely used graphical techniques:

- Box plots
- Histograms
- Normal probability plots
- Scatter plots

Box plots

You use box plots to show some of the most important features of a dataset, such as the following:

- ✔ Minimum value
- ✔ Maximum value
- ✔ Quartiles

Quartiles separate a dataset into four equal sections. The first quartile (Q_1) is a value such that the following is true:

25 percent of the observations in a dataset are less than the first quartile.

75 percent of the observations are greater than the first quartile.

The second quartile (Q_2) is a value such that

50 percent of the observations in a dataset are less than the second quartile.

50 percent of the observations are greater than the second quartile.

The second quartile is also known as the *median*.

The third quartile (Q_3) is a value such that

75 percent of the observations in a dataset are less than the third quartile.

25 percent of the observations are greater than the third quartile.

You can also use box plots to identify *outliers*. These are values that are substantially different from the rest of the dataset. Outliers (the topic of Chapter 10) can cause problems for traditional statistical tests, so it's important to identify them before performing any type of statistical analysis.

Histograms

You use histograms to gain insight into the probability distribution that a dataset follows. With a histogram, the dataset is organized into a series of individual values or ranges of values, each represented by a vertical bar. The height of the bar shows how frequently a value or range of values occurs. With a histogram, it's easy to see how the data is distributed.

Scatter plots

A scatter plot is a series of points that show how two variables are related to each other. A random scatter of points indicates that the two variables are unrelated, or that the relationship between them is very weak. If the points closely resemble a straight line, this indicates that the relationship between the two variables is approximately *linear*.

Two variables are linearly related if they can be described with the equation $Y = mX + b$.

X is the independent variable, and Y is the dependent variable. m is the *slope,* which represents the change in Y due to a given change in X. b is the *intercept,* which shows the value of Y when X equals zero.

Figure 11-1 shows a scatter plot between two variables in which the relationship appears to be linear.

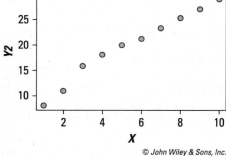

Figure 11-1:
Scatter plot
of a linear
relationship.

The points on the scatter plot in Figure 11-1 very nearly form a straight line. It bends a little to the left and bends a little to the right, but it's roughly straight. This shows that the relationship is linear, with a positive slope.

Figure 11-2 shows a scatter plot between two variables in which Y appears to be rising more rapidly than X.

See the curve in Figure 11-2? This relationship is clearly not linear. It is in fact a quadratic relationship. A quadratic relationship takes the form
$Y = aX^2 + bX + c.$

Figure 11-3 shows a scatter plot in which there doesn't appear to be any relationship between X and Y.

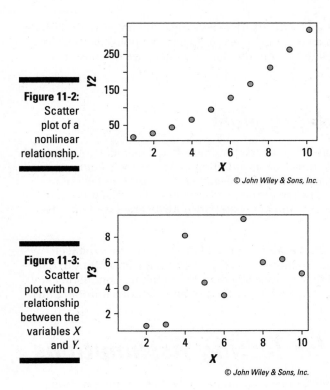

Figure 11-2:
Scatter
plot of a
nonlinear
relationship.

© John Wiley & Sons, Inc.

Figure 11-3:
Scatter
plot with no
relationship
between the
variables *X*
and *Y*.

© John Wiley & Sons, Inc.

The variables in the scatter plot shown in Figure 11-3 are *unrelated* or *independent;* you can see this by the lack of any pattern in the data.

In addition to showing the relationship between two variables, a scatter plot can also show the presence of outliers. Figure 11-4 shows a dataset with one observation that is substantially different from the other observations.

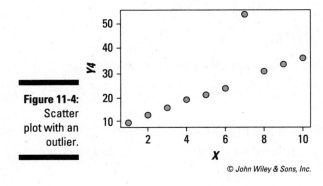

Figure 11-4:
Scatter
plot with an
outlier.

© John Wiley & Sons, Inc.

The outlier point in Figure 11-4 needs to be investigated further to determine whether it's the result of an error or other problems. It's possible that the outlier will need to be removed from the data. (Outliers are discussed in detail in Chapter 10.)

Normal probability plots

Normal probability plots are used to see how closely the elements of a dataset follow the normal distribution. The assumption of normality is common in many disciplines. For example, it's often assumed in finance and economics that the returns to stocks are normally distributed. The assumption of normality is very convenient, and many statistical tests are based on this assumption.

Applying statistical tests that assume normality to a *non-normal* dataset would give extremely questionable results. Therefore, it's important to determine whether or not the data is normally distributed before conducting any of these statistical tests. Chapter 8 talks about some methods for doing this.

EDA Techniques for Testing Assumptions

There are several EDA techniques you can use to test assumptions about a dataset. These include

- Run sequence plot
- Lag plot
- Histogram
- Normal probability plot

Run sequence plot

Many statistical techniques are based on the assumption that the data being analyzed has the following properties:

- Independent variables
- Variables drawn from a common probability distribution
- Variables with common parameters (for example, mean and standard deviation)

A *run sequence plot* tests whether the data conforms to these assumptions. For example, Figure 11-5 shows a run sequence plot for the daily returns to the Standard and Poor's stock market index.

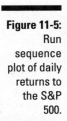
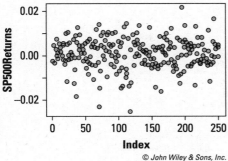

Figure 11-5: Run sequence plot of daily returns to the S&P 500.

© *John Wiley & Sons, Inc.*

Because this is a time series plot, it's being used to determine whether the returns to the S&P 500 are independent of each other, whether they are all drawn from the same probability distribution, and whether the parameters (mean and variance) remain constant over time.

The run sequence plot is designed to answer these questions:

✔ Are there any changes in the mean of the data?

✔ Are there any changes in the variance of the data?

In addition, you use the run sequence plot to identify any outliers in the data.

The plot of the returns to the S&P 500 shows that the mean and variance of the data remain stable over time, and that there do not appear to be any outliers.

Lag plot

A *lag plot* determines whether the elements of a dataset are *random* (independent of each other). In other words, the plot shows whether or not there's a pattern in the data. Patterns in the data are inconsistent with randomness.

A lagged value is one that has occurred in the past. A lag of 1 refers to an observation that has taken place one period in the past. A lag of 2 refers to an observation that has taken place two periods in the past, and so forth.

A lag plot shows the values of a variable on the vertical axis, and the lagged values of the same variable on the horizontal axis. For example, Figure 11-6 shows a lag plot for the daily returns to the Standard and Poor's stock market index.

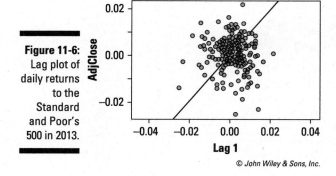

Figure 11-6:
Lag plot of daily returns to the Standard and Poor's 500 in 2013.

© John Wiley & Sons, Inc.

The points on this plot are randomly scattered with no particular pattern. This is consistent with the assumption of randomness in the data.

Histogram

You can use a *histogram* to identify the distribution followed by a dataset. A histogram can show several key details about a dataset, including the following:

- ✔ The center of the data
- ✔ The spread (variability) of the data
- ✔ The skewness of the data (if any)
- ✔ The presence of outliers

For example, Figure 11-7 shows a histogram for the daily returns to the Standard and Poor's stock market index.

The graph shows that the Standard and Poor's returns have a mean of approximately 0 — the heights of the bars are greatest near 0. The returns appear to exhibit *negative skewness* (that is, extreme negative returns are more common than extreme positive returns) and have a greater magnitude. There do not appear to be any outliers in the data.

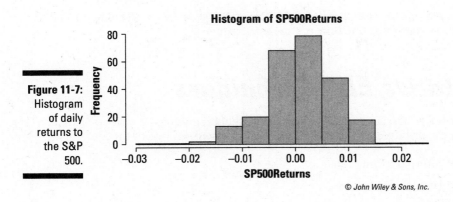

Figure 11-7:
Histogram
of daily
returns to
the S&P
500.

© John Wiley & Sons, Inc.

Normal probability plot

Use a *normal probability plot* to compare a dataset to the normal distribution. The vertical axis of this plot shows the quantiles of the dataset, and the horizontal axis shows the quantiles of the normal distribution. If a dataset is normally distributed, then the graph should appear to be a straight line with a slope of 1.

Quantiles are used to divide a dataset into equally sized groups. A widely used type of quantile is the *quartile,* which (as discussed earlier) divides a dataset into four equal groups, each consisting of 25 percent of the data. Another popular choice is the *percentile,* which divides a dataset into one hundred equal groups, each consisting of 1 percent of the data.

Figure 11-8 shows a normal probability plot for the daily returns to the Standard and Poor's stock market index.

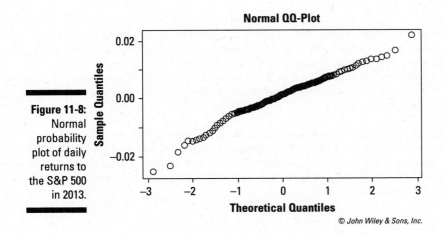

Figure 11-8:
Normal
probability
plot of daily
returns to
the S&P 500
in 2013.

© John Wiley & Sons, Inc.

The plot shows that the returns to the S&P 500 are close to being normal, with deviations in the tails of the distribution.

Quantitative EDA Techniques

Although EDA is mainly based on graphical techniques, it also consists of a few quantitative techniques. This section discusses two of these:

- ✔ Interval estimation
- ✔ Hypothesis testing

Interval estimation

Interval estimation is a technique that's used to construct a *range* of values within which a variable is likely to fall. One important example of this is the confidence interval. A *confidence interval* is a range of numbers that is likely to contain the value of a population measure such as the mean. A confidence interval is constructed as follows:

Point estimate ± Margin of error

The confidence interval consists of a *lower limit* equal to the point estimate *minus* the margin of error, and an *upper limit* equal to the point estimate *plus* the margin of error.

The *point estimate* is a single value estimated from a sample. For example, the sample mean is a point estimate of the population mean. Similarly, the sample standard deviation is a point estimate of the population standard deviation.

The margin of error reflects the amount of uncertainty associated with the point estimate. In other words, it shows how much the point estimate can change from one sample to the next. The margin of error is based on the standard deviation and the size of the sample being used. The result of these calculations is a range of values that is likely to contain the true value of the population measure.

For example, suppose a researcher determines that with 95 percent confidence, the interval (–2.0 percent, +8.0 percent) contains the true value of the mean return to the S&P 500 next year. The sample mean is the average of the lower and upper limit of this interval (that is, 3.0 percent). The margin of error is therefore 5 percent.

Hypothesis testing

A *statistical hypothesis* is a statement that's assumed to be true unless there's strong contradictory evidence. (Chapter 5 introduces hypothesis testing, and Chapters 8, 10, and 14 talk about it extensively.) Hypothesis testing is widely used in many disciplines to determine whether a proposition is true or false. For example, hypothesis testing could be used to determine whether

✔ The mean age of the residents of a state is 43 years old.

✔ The mean return to the stocks in a portfolio is 7.2 percent.

✔ The amount of annual rainfall in a city follows the normal distribution.

Hypothesis testing is a multi-step process consisting of the following:

1. **The statement of the null hypothesis:** This is the statement that is assumed to be true.

2. **The statement of the alternative hypothesis:** This is the statement that will be accepted if the null hypothesis is rejected.

3. **The level of significance at which the hypothesis test will be conducted:** This equals the likelihood of rejecting the null hypothesis when it is false.

4. **The test statistic:** This is a numerical measure that shows whether sample data is consistent with the null hypothesis.

5. **The critical value:** If the test statistic is more extreme than the critical value, the null hypothesis is rejected.

6. **The decision:** Based on the relationship between the test statistic and the critical value, you make a decision as to whether or not the null hypothesis should be rejected.

Chapter 12

A Plot to Get Graphical: Graphical Techniques

In This Chapter

▶ Delving into several graphical techniques you can use to analyze data

▶ Comprehending the advantages and disadvantages of different plotting approaches

▶ Determining how to choose the right graphical technique for any situation

---

*Y*ou use graphs to gain insights into the statistical properties of a dataset. It's often easy to see patterns or relationships in a dataset that would be easy to miss with numerical techniques, and this is especially true for large datasets.

This chapter introduces several of the most important graphical techniques for analyzing datasets:

- ✔ Stem-and-leaf plots
- ✔ Scatter plots
- ✔ Box plots
- ✔ Histograms
- ✔ Quantile-quantile (QQ) plots
- ✔ Autocorrelation plots

Stem-and-Leaf Plots

A *stem-and-leaf plot* is a graphical device in which the distribution of a dataset is organized by the numerical value of the observations in the dataset. The diagram consists of a "stem," showing the different categories in the data, and a "leaf," which shows the values of the individual observations in the dataset.

For example, the following is a stem-and-leaf diagram for the daily prices of Microsoft stock from January 1, 2013 to December 31, 2013. The prices range from $25.16 to $38.14:

```
25 | 22444567899999
26 | 0111222333455555666677777788889999
27 | 00000001222444455666678
28 | 45
29 | 036
30 | 333344555556666777788899999
31 | 00333345556666777888889999
32 | 01111123334444556667788888999
33 | 0001122222234445555666677778888888889
34 | 000224444555566799
35 | 0568899
36 | 0112333555556667788889
37 | 0112345679
38 | 1
```

On the stem-and-leaf plot, each line represents a single category; for this dataset, each category is a dollar amount. For example, the category 32 consists of all prices between $32.00 and $32.99. Each price for Microsoft stock is quoted in dollars and cents. The left side of the bar shows the dollars (the stems); the right side of the bar shows the cents (the leaves), after rounding to the nearest 10 cents. For example, a price of $32.23 is rounded to $32.20, and this appears as a 2 on the right-hand side of the bar for the category 32. A price of $33.48 is rounded to $33.50; this appears as a 5 on the right-hand side of the bar for the category 33.

Using this technique, it's easy to see how many prices fall into each category. For example, there were 14 trading days in the dataset in which the price of Microsoft stock was between $25.00 and $25.99. There were three trading days in which the price of Microsoft stock was between $29.00 and $29.99. A price between $33.00 and $33.99 occurred most frequently, and a price between $38.00 and $38.99 was the most infrequent during the year.

One of the advantages of a stem-and-leaf diagram is that it's easy to identify the *mode* of a dataset. (Recall that the mode is the value that occurs most frequently in a dataset.) If you look only at the dollar ranges, then it's easy to spot which range contains the most observations — the one with the longest leaf. In this case, a price in the 33 range ($33.00–$33.99) would be considered to be the mode because it contains the most observations.

Another advantage of this diagram is that *outliers* are easy to spot. An outlier is an observation in a dataset that is significantly larger or smaller than the other observations in the dataset. An outlier would be indicated by a large gap between either the first or last stem and the next closest one. (Chapter 10 talks more about outliers.)

One drawback to stem-and-leaf diagrams is that they become difficult to interpret for large datasets because the size of the leaf becomes unwieldy.

Scatter Plots

Unlike a stem-and-leaf plot, a *scatter plot* is intended to show the relationship between *two* variables. It may be difficult to see whether there's a relationship between two variables just by looking at the raw data, but with a scatter plot, any patterns that exist in the data become much easier to see.

A scatter plot consists of a series of points; each point shows a single value for two different variables. For example, you could construct a scatter plot to show the relationship between a corporation's annual revenues and its annual profits. If you're trying to predict profits based on revenue, the X-axis would be used to show annual revenues, and the Y-axis would be used to show annual profits. So, revenues are considered to be the independent variable, and profits are considered to be the dependent variable in this relationship.

On a scatter plot, the X-axis (that is, the horizontal axis) is used to show the *independent* variable, and the Y-axis (the vertical axis) is used to show the *dependent* variable.

In this example, each point on the scatter plot shows the revenues and sales for a specified year. Table 12-1 shows the relationship between the annual revenues and annual profits of a corporation during the years 2004–2014.

Figure 12-1 shows the resulting scatter plot.

Each point on the scatter plot in Figure 12-1 represents the revenues and related profits of the corporation for a single year. For example, the point in the upper right-hand corner of the plot represents data for 2014, when the corporation's revenues were $312 million and the profits were $71 million.

Table 12-1	Annual Revenues and Profits 2004–2014	
Year	Revenues ($ millions)	Profits ($ millions)
2004	225	42
2005	237	43
2006	245	48
2007	222	40
2008	265	60
2009	270	56
2010	254	53
2011	280	60
2012	290	62
2013	305	65
2014	312	71

Figure 12-1:
Scatter plot
of annual
revenues
and profits
2004–2014.

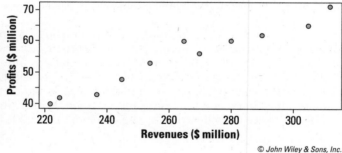

© John Wiley & Sons, Inc.

You can use a scatter plot to determine whether

- The two variables tend to move in the same direction.
- The two variables tend to move in opposite directions.
- The two variables aren't related to each other.

If two variables tend to move in the same direction, they are said to be *positively correlated;* if they tend to move in opposite directions, they are said to be *negatively* correlated. If two variables don't show any particular pattern, they are said to be *uncorrelated.*

Figure 12-2 shows a scatter plot for two variables (X and Y) that are positively correlated.

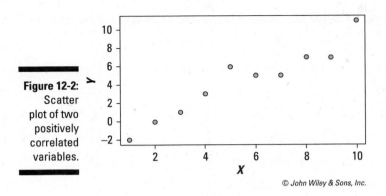

Figure 12-2:
Scatter plot of two positively correlated variables.

The scatter plot shows that as X increases, there's a strong tendency for Y to increase (but not necessarily by the same amount). This shows that X and Y are positively correlated.

Figure 12-3 shows the same scatter plot with a *trend line;* the equation of this line is estimated with *regression analysis.* (See Chapter 15 for a discussion of how to implement regression analysis and interpret the results.)

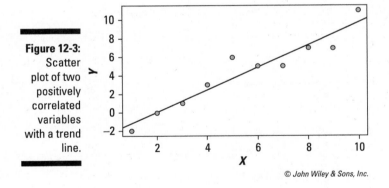

Figure 12-3:
Scatter plot of two positively correlated variables with a trend line.

The trend line shows how much Y changes on average, given a specific change in X. A positively sloped trend line indicates that two variables are positively correlated; similarly, a negatively sloped trend line indicates that two variables are negatively correlated. If a trend line is flat (that is, has a zero slope), this indicates that the two variables are unrelated to each other. In Figure 12-3, the positively sloped trend line shows that X and Y are positively correlated.

Figure 12-4 shows a scatter plot and the trend line for two variables that are *negatively* correlated.

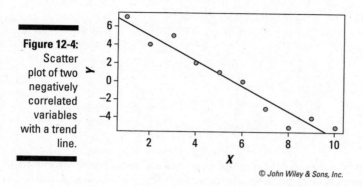

Figure 12-4: Scatter plot of two negatively correlated variables with a trend line.

The scatter plot shows that as X increases, Y tends to decrease; the trend line has a *negative* slope. Therefore, X and Y are *negatively* correlated.

Figure 12-5 shows a scatter plot and the trend line for two variables that are *uncorrelated*.

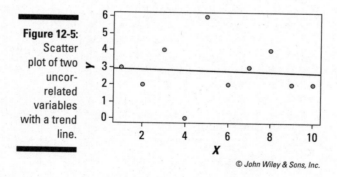

Figure 12-5: Scatter plot of two uncorrelated variables with a trend line.

The scatter plot shows that as X increases, Y sometimes increases and sometimes decreases. There is no particular pattern to the data. The points appear to be randomly scattered throughout the diagram. As a result, the trend line is nearly flat, and this shows that X and Y are *uncorrelated*.

For a real-world example, Figure 12-6 shows a scatter plot of the relationship between the price of Apple stock and the Standard and Poor's 500 stock market index from January 1, 2013 to December 31, 2013.

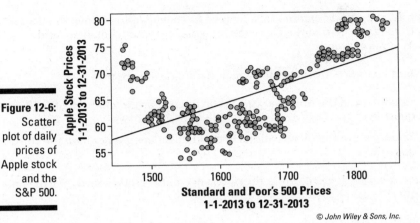

Figure 12-6:
Scatter plot of daily prices of Apple stock and the S&P 500.

© John Wiley & Sons, Inc.

The Standard and Poor's 500 (S&P 500) index is an average of the prices of the 500 largest U.S. stocks. The value of this index reflects the overall state of the U.S. economy. The plot shows that there's a positive correlation between the price of Apple stock and the S&P 500.

Box Plots

A *box plot* is designed to show several key statistics for a dataset in the form of a vertical rectangle or box. The statistics it can show include the following:

- Minimum value
- Maximum value
- First quartile (Q_1)
- Second quartile (Q_2)
- Third quartile (Q_3)
- Interquartile range (IQR)

The *first quartile* of a dataset is a numerical measure that divides the data into two parts: the smallest 25 percent of the observations and the largest 75 percent of the observations. In other words, the first quartile is a numerical value with the following properties:

- 25 percent of the observations in the dataset are *smaller than* the first quartile.
- 75 percent of the observations in the dataset are *greater than* the first quartile.

Similarly, the *second quartile* (also known as the *median*) divides the data in half, so 50 percent of the elements are smaller than the median, and 50 percent are larger.

The *third quartile* is the value for which the following are true:

✔ 75 percent of the observations in the dataset are *smaller than* the third quartile.

✔ 25 percent of the observations in the dataset are *greater than* the third quartile.

The *interquartile range (IQR)* is the difference between the third quartile and first quartile: $IQR = Q_3 - Q_1$.

The interquartile range is a measure of *dispersion;* it shows how much spread there is between the elements in the middle 50 percent of a dataset.

A box plot is drawn so that

✔ The top of the box represents the *third quartile* (Q_3) of the data.

✔ The bottom of the box represents the *first quartile* (Q_1) of the data.

✔ The middle of the box (shown with a line) represents the *second quartile* (Q_2).

In addition, there's a line *above* the box to indicate the *maximum* value in the data that doesn't exceed $Q_3 + 1.5 \times IQR$ and a line *below* the box to indicate the *minimum* value in the data that doesn't fall below $Q_1 - 1.5 \times IQR$. Values outside of this range are *outliers* and are shown on the box plot as individual points. (Outliers are discussed in Chapter 10.)

Figure 12-7 shows a box plot of the daily prices of Microsoft stock from January 1, 2013 to December 31, 2013.

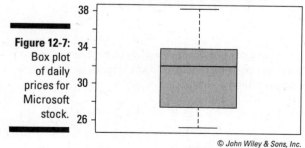

Box Plot of Microsoft Prices

Figure 12-7: Box plot of daily prices for Microsoft stock.

There are no outliers in this data. Therefore, the bottom line in the box plot shows that the lowest price during this period was somewhat less than $26.00, and the top line shows that the highest price was just over $38. The bottom of the box corresponds to the first quartile, which is $27.43; the solid line in the middle of the box corresponds to the second quartile (median), which is $31.89. The top of the box corresponds to the third quartile, which is $33.78. The height of the box equals the interquartile range (IQR), which is $6.35.

As another example, Figure 12-8 shows a box plot of the daily prices of Apple stock from January 1, 2013 to December 31, 2013.

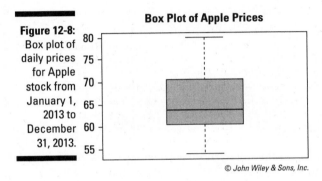

Figure 12-8: Box plot of daily prices for Apple stock from January 1, 2013 to December 31, 2013.

Box Plot of Apple Prices

© John Wiley & Sons, Inc.

The lowest price in 2013 for Apple stock was $53.84, and the highest price was $80.11. There are no outliers in the data, so these values are shown by the bottom line and top line, respectively.

The first quartile, shown at the bottom of the box, was $60.48. The second quartile was $63.65 (shown by the solid black line) and the third quartile was $70.32, shown at the top of the box. As a result, the interquartile range (IQR) is $9.84.

Histograms

A *histogram* is a graph that represents the probability distribution of a dataset. A histogram has a series of vertical bars where each bar represents a single value or a range of values for a variable. The heights of the bars indicate the frequencies or probabilities for the different values or ranges of values.

For example, Figure 12-9 shows a histogram of the daily prices of Apple stock from January 1, 2013 to December 31, 2013.

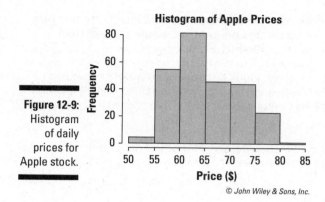

Figure 12-9:
Histogram
of daily
prices for
Apple stock.

© John Wiley & Sons, Inc.

According to this histogram, most of the prices were between $60 and $65; the price was in this range 81 times during the year. The second most frequently observed prices were between $55 and $60; the price landed in this range 44 times during the year. The third most frequent range of prices was between $65 and $70, and the fourth most frequent range of prices was between $70 and $75. Very few prices were between $50 and $55, and the fewest prices observed during the year were between $80 and $85.

Based on the graph, the mean and median price were close to the $60 to $65 range. The actual mean was $65.67, and the actual median was $63.65. Since the mean exceeds the median, the distribution of prices for 2013 was *positively skewed.* This indicates that the likelihood of an extremely large price is somewhat greater than the likelihood of an extremely low price.

A distribution is positively skewed if the mean is greater than the median; it is negatively skewed if the mean is less than the median. The distribution is symmetrical about the mean if the mean equals the median. How much the data is skewed depends on how far the mean and median differ. If they are very close, it's sometimes practical to treat the distribution as symmetric.

As another example, Figure 12-10 shows a histogram of the daily prices of the S&P 500 stock index from January 1, 2013 to December 31, 2013.

According to the histogram in Figure 12-10, the most frequently observed range of prices during the year was between $1,650 and $1,700. The mean turned out to be $1,643.80, and the median was $1,650.41. Unlike Apple stock, the mean was *below* the median; the distribution of prices for 2013 is *negatively skewed.* This indicates that there was a slightly greater tendency for the Standard and Poor's 500 to trade below the mean than above the mean in 2013.

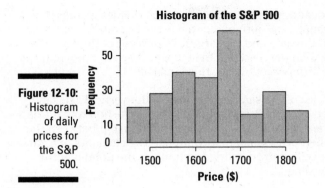

Histogram of the S&P 500

Figure 12-10:
Histogram
of daily
prices for
the S&P
500.

One of the most important uses of histograms is to determine if a dataset follows a specified probability distribution. Although there are many formal statistical tests to determine which probability distribution a dataset follows, it's good practice to visually inspect the data with a graph before engaging in any formal statistical tests.

Figure 12-9 provides strong evidence that Apple stock prices are *not* normally distributed. The normal distribution is *symmetrical* about its mean, whereas the Apple stock prices are positively skewed. Figure 12-10 provides strong evidence that the S&P 500 is also unlikely to be normally distributed because its distribution is negatively skewed.

Formal statistical tests would be required to show that neither distribution is normal, but the graphs are highly suggestive. Because many statistical tests are based on the assumption of normality, it's important to determine if a distribution is truly normal before you use any of these tests.

Quantile-Quantile (QQ) Plots

A *quantile-quantile plot* (also known as a *QQ-plot*) is another way you can determine whether a dataset matches a specified probability distribution. QQ-plots are often used to determine whether a dataset is *normally* distributed. Graphically, the QQ-plot is very different from a histogram. As the name suggests, the horizontal and vertical axes of a QQ-plot are used to show *quantiles*.

We used quartiles earlier in this chapter to create stem-and-leaf plots. Quartiles divide a dataset into four equal groups, each consisting of 25 percent of the data. But there is nothing particularly special about the number four. You can choose any number of groups you please.

Another popular type of quantile is the *percentile,* which divides a dataset into 100 equal groups. For example, the 30th percentile is the boundary between the smallest 30 percent of the data and the largest 70 percent of the data. The median of a dataset is the 50th percentile of the dataset. The 25th percentile is the first quartile, and the 75th percentile the third quartile.

With a QQ-plot, the quantiles of the sample data are on the vertical axis, and the quantiles of a specified probability distribution are on the horizontal axis. The plot consists of a series of points that show the relationship between the actual data and the specified probability distribution. If the elements of a dataset perfectly match the specified probability distribution, the points on the graph will form a 45 degree line.

For example, Figure 12-11 shows a normal QQ-plot for the price of Apple stock from January 1, 2013 to December 31, 2013.

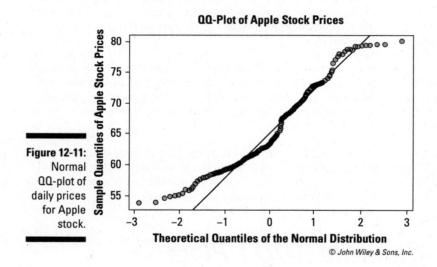

QQ-Plot of Apple Stock Prices

Figure 12-11:
Normal
QQ-plot of
daily prices
for Apple
stock.

© *John Wiley & Sons, Inc.*

The QQ-plot shows that the prices of Apple stock do not conform very well to the normal distribution. In particular, the deviation between Apple stock prices and the normal distribution seems to be greatest in the lower left-hand corner of the graph, which corresponds to the *left tail* of the normal distribution. The discrepancy is also noticeable in the upper right-hand corner of the graph, which corresponds to the *right tail* of the normal distribution.

The graph shows that the smallest prices of Apple stock are not small enough to be consistent with the normal distribution; similarly, the largest prices of Apple stock are not large enough to be consistent with the normal distribution. This shows that the tails of the Apple stock price distribution are too "thin" or "skinny" compared with the normal distribution. The conclusion to be drawn from this is that the Apple stock prices are *not* normally distributed.

Figure 12-12 shows a normal QQ-plot for the daily *returns* to Apple stock from January 1, 2013 to December 31, 2013:

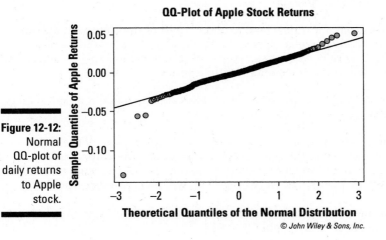

Figure 12-12:
Normal
QQ-plot of
daily returns
to Apple
stock.

© John Wiley & Sons, Inc.

The QQ-plot shows that the returns to Apple stock do not conform to the normal distribution, either. In this case, the smallest returns to Apple stock are too small to be consistent with the normal distribution. Similarly, the largest returns to Apple stock are too large to be consistent with the normal distribution. This shows that the tails of the Apple return distribution are too "thick" or "fat" compared with the normal distribution. Therefore, Apple returns are *not* normally distributed.

In many applications, the returns to financial assets are assumed to be normally distributed, but in actual practice, these returns tend to have "fat" tails. With a fat-tailed distribution, extremely large or small outcomes occur more frequently than they would with the normal distribution. As Chapter 15 points out, there are ways of transforming the data to bring it more in line with the normal distribution.

Autocorrelation Plots

An *autocorrelation plot* shows the properties of a type of data known as a time series. (Chapter 16 discusses time series analysis in detail.)

A *time series* refers to observations of a single variable over a specified time horizon. For example, the daily price of Microsoft stock during the year 2013 is a time series. *Cross-sectional data* refers to observations on many variables at a single point in time. For example, the closing prices of the 30 stocks contained in the Dow Jones Industrial Average on January 31, 2014, would be considered cross-sectional data.

An autocorrelation plot is designed to show whether the elements of a time series are positively correlated, negatively correlated, or independent of each other. (The prefix *auto* means "self"— autocorrelation specifically refers to correlation among the elements of a time series.)

An autocorrelation plot shows the value of the autocorrelation function (acf) on the vertical axis. It can range from –1 to 1.

The horizontal axis of an autocorrelation plot shows the size of the *lag* between the elements of the time series. For example, the autocorrelation with lag 2 is the correlation between the time series elements and the corresponding elements that were observed two time periods earlier.

Figure 12-13 shows an autocorrelation plot for the daily prices of Apple stock from January 1, 2013 to December 31, 2013.

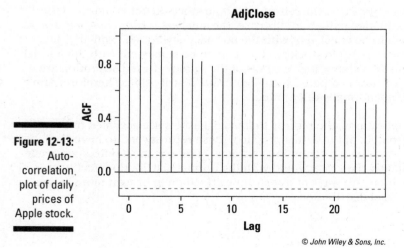

Figure 12-13: Auto-correlation plot of daily prices of Apple stock.

© John Wiley & Sons, Inc.

On the graph, there is a vertical line (a "spike") corresponding to each lag. The height of each spike shows the value of the autocorrelation function for the lag.

The autocorrelation with lag zero always equals 1, because this represents the autocorrelation between each term and itself. Price and price with lag zero are the same variable.

Each spike that rises above or falls below the dashed lines is considered to be *statistically significant*. (Chapter 16 talks about this in detail.) This means the spike has a value that is significantly different from zero. If a spike is significantly different from zero, that is evidence of autocorrelation. A spike that's close to zero is evidence against autocorrelation.

In this example, the spikes are statistically significant for lags up to 24. This means that the Apple stock prices are highly correlated with each other. In other words, when the price of Apple stock rises, it tends to continue rising. When the price of Apple stock falls, it tends to continue falling. Figure 12-14 illustrates this.

Time Series Plot of Apple Stock Prices

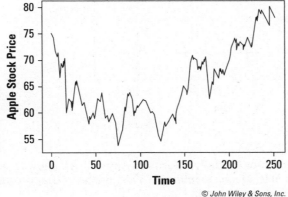

Figure 12-14:
Time series plot of daily prices of Apple stock.

© John Wiley & Sons, Inc.

Even though the daily prices of Apple stock are highly correlated, the daily returns may not be. You compute the daily returns from the daily prices as follows:

$$r_t = \ln\left(\frac{P_t}{P_{t-1}}\right)$$

where

r_t = The *continuously compounded* return at time t

P_t = The price at time t

P_{t-1} = The price at time $t - 1$ (one period before t)

ln = The natural logarithm

The natural logarithm is the logarithm with base e, which is approximately equal to 2.71828. . . .

Figure 12-15 shows an autocorrelation plot for the daily returns to Apple stock from January 1, 2013 to December 31, 2013.

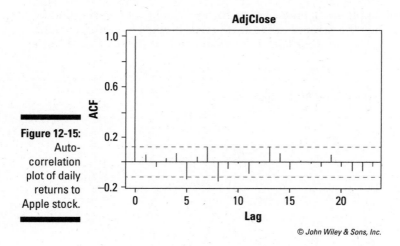

Figure 12-15:
Auto-
correlation
plot of daily
returns to
Apple stock.

© *John Wiley & Sons, Inc.*

The autocorrelation plot for daily returns to Apple stock shows that most of the spikes are not statistically significant. This indicates that the returns are not highly correlated, as shown in Figure 12-16.

The graph shows that except for one major downturn, the returns to Apple stock between January 1, 2013 and December 31, 2013 do not show any particular pattern — they tend to fluctuate randomly around zero. This means that the returns are largely independent of each other.

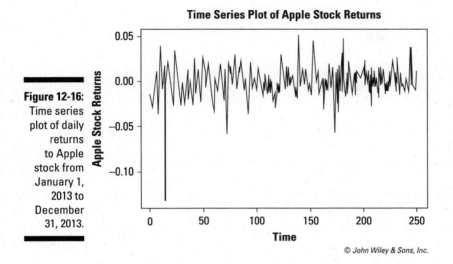

Figure 12-16: Time series plot of daily returns to Apple stock from January 1, 2013 to December 31, 2013.

You can use an autocorrelation plot to determine whether the elements of a time series are *random* (that is, unrelated to each other). This is important, because many statistical tests involving time series are based on this assumption.

As you can see, there are many different ways to visualize your data. A picture is worth a thousand words, as the saying goes. And it definitely holds true in data analysis. Statistical software packages generally come equipped with easy-to-use graphical tools. By taking advantage of them, you can quickly gain insight into your data that no amount of number crunching could give you.

Chapter 13

You're the Only Variable for Me: Univariate Statistical Techniques

Chapter 4 introduces the notion of a *probability distribution*. You use a probability distribution to describe the properties of a random variable. A *random variable* is actually a function — it assigns numerical values to the outcomes of a random process. For example, you might define a random variable as "the return to Apple stock over the coming trading day" or "the number of bonds in a portfolio that default over the coming year."

The two basic types of probability distributions you may encounter in statistics are discrete distributions and continuous distributions:

✓ A *discrete distribution* is one for which every outcome has a positive probability.

✓ A *continuous distribution* assigns probabilities to ranges of values.

For example, the distribution of bonds that default over the coming year would be discrete, because every bond has a distinct probability of defaulting. The distribution of returns to Apple stock is continuous because we assign to probabilities to the return falling into ranges. We talk about the probability that the return will exceed 10%, for example.

Each probability distribution is appropriate for different types of applications. For example, you can model the likelihood of bond defaults over the coming year based on the *binomial* distribution, because a default is an all-or-nothing event. (Recall from Chapter 4 that you use the binomial distribution to describe situations where only one of two things can happen.) You can model the returns to Apple stock based on the *normal* distribution because the normal distribution is a continuous distribution with several convenient properties.

The discrete distributions covered in this chapter are the binomial and the Poisson. Both are widely used in business applications. The chapter also covers several continuous distributions that are prevalent in business applications, such as the normal, Student's t-distribution, lognormal, chi-square, and F-distribution.

Each of the distributions in this chapter is an example of a *univariate* distribution. That is, these distributions describe the properties of a single random variable. By contrast, a *multivariate* distribution describes the joint behavior of two or more random variables. (For more on multivariate distributions, see Chapter 14.)

This chapter also discusses statistical tests of randomness, along with methods for determining whether any data is an outlier. An *outlier* is a value that's substantially different from other values in the same sample; in some cases, you need to remove outliers from a dataset to avoid distorting the estimation results.

Counting Events Over a Time Interval: The Poisson Distribution

The *Poisson distribution* is useful for measuring how many events may occur during a given time horizon, such as the number of customers who enter a store during the next hour, the number of hits a website gets during the next minute, and so forth. The Poisson distribution is based on the *Poisson process,* where events are counted over intervals of time. What characterizes a Poisson process is that the number of events observed in a given time period does not depend at all on the number of events that were observed in the previous time period (or any other time period).

One type of situation that tends to follow a Poisson distribution is waiting times for a bus or subway train. It turns out that your expected waiting time for the next bus remains the same no matter how long you've already been waiting. This is the signature (and rather strange) feature of the Poisson distribution. The fact that you've been waiting for 20 minutes has no bearing on the probability that a bus will arrive in the next 5 minutes.

Computing Poisson probabilities

You calculate Poisson probabilities with the following formula:

$$P(X = x) = e^{-\lambda} \frac{\lambda^x}{x!}$$

where

X = A Poisson random variable

x = Number of events that occur

λ = The Greek letter *lambda,* used to represent the average number of events that occur per time interval. Obviously, the value of λ depends on length of the time interval you choose.

e = A *constant* that is approximately equal to 2.71828

For example, suppose a bank discovers that from historical experience, on average, three customers use the ATM in its lobby each hour. What is the probability that during the next hour, five customers will use the ATM?

In this case, the value of lambda (λ) is equal to 3 because the average number of customers using the ATM each hour equals 3. The probability of five customers using the ATM during the next hour is computed using the Poisson formula, with $\lambda = 3$ and $x = 5$:

$$P(X = 5) = e^{-3} \frac{3^5}{5!} = 0.1008$$

If you don't care for using formulas or a table, try a specialized calculator or Excel. For Excel 2007 and older versions, use the POISSON function; for Excel 2010, use the POISSON.DIST function.

The values you must supply are x (the number of events), λ, and a variable known as cumulative. This value equals 1 if you want a "less than or equal to" probability; otherwise, this value equals 0. Based on the ATM example, you would enter =POISSON.DIST(5, 3, 0).

Moments of the Poisson distribution

Just like the binomial distribution, the Poisson distribution has specialized formulas for computing the expected value, variance, and standard deviation. These appear in the following sections.

Poisson distribution: Calculating the expected value

The expected value of the Poisson distribution is computed like this:

$$E(X) = \lambda$$

This is because λ represents the *average* number of events that occur during a given time interval. In the ATM example, the average number of customers who use the ATM per hour is three. That's the expected value of X.

Poisson distribution: Computing variance and standard deviation

Compute the variance of the Poisson distribution as $\sigma^2 = \lambda$; the standard deviation (σ) equals the square root of λ. For the ATM example, the variance is 3, and the standard deviation is the square root of 3, which is approximately 1.732.

Graphing the Poisson distribution

As with the binomial distribution, the Poisson distribution can be illustrated with a histogram. For the ATM example, $\lambda = 3$. Figure 13-1 shows the histogram of the Poisson distribution with $\lambda = 3$.

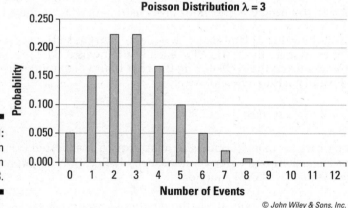

Figure 13-1:
Poisson
distribution
with $\lambda = 3$.

© John Wiley & Sons, Inc.

Continuous Probability Distributions

As discussed in Chapter 4, continuous probability distributions are used in situations where probabilities are assigned to ranges of values rather than to specific outcomes. Some of the most widely used continuous distributions in business applications are the normal, Student's t, lognormal, chi-square, and F-distributions. The normal distribution is introduced in detail in Chapter 4. We discuss the others here.

The Student's t-distribution

The Student's t-distribution (also known simply as the Student's t or simply the t-distribution) was derived in the early 20th century by William S. Gosset, who was employed as a statistician at the Guinness Brewing Company and wrote about it under the pseudonym *Student*. The t-distribution was developed to describe the statistical properties of sample means that are estimated from *small* samples. As a rule of thumb, "small" generally means less than 30.

You will run across the Student's t-distribution several times in the ensuing chapters. It is useful in several contexts related to evaluating predictive models. For now, this section concentrates on describing the properties of the Student's t.

Comparing the t-distribution with the standard normal distribution

The t-distribution has several important properties in common with the standard normal distribution:

- ✔ It has a mean of zero.
- ✔ It is *symmetrical* about the mean — that is, the area below the mean is a mirror image of the area above the mean.
- ✔ It can be described graphically with a *bell-shaped curve*.

The main *difference* between the two distributions is that the variance and standard deviation of the t-distribution are larger than those of the standard normal distribution. As a result, the t-distribution has more area in the "tails" and less area near the mean than the standard normal distribution. (For this reason, the t-distribution is sometimes said to have "fat tails.")

The variance and standard deviation of the standard normal distribution are both equal to 1. The mean equals 0.

The larger variance and standard deviation in the t-distribution reflect the fact that there will be much more *variability* among the means of small samples than among the means of large samples.

The t-distribution is actually an infinite set of distributions, each uniquely characterized by the degrees of freedom. The larger the number of degrees of freedom, the more closely the t-distribution resembles the standard normal distribution. With 30 or more degrees of freedom, the two distributions are extremely similar to each other. This is why only the first 30 of these distributions is widely used in applications.

Figure 13-2 shows the standard normal distribution and the t-distribution with two degrees of freedom.

The graph shows that the t-distribution has more area in the tails and less area around the mean than the standard normal distribution does. As a result, more extreme observations (positive and negative) are likely to occur under the t-distribution than under the standard normal distribution. The t-distribution reflects the fat tails referenced earlier.

With 30 degrees of freedom, the t-distribution and the standard normal distribution are almost indistinguishable. You can see this in Figure 13-3.

Moments of the t-distribution

For the t-distribution with v degrees of freedom, the mean (or expected value) equals the following:

$$\mu = E(X) = 0$$

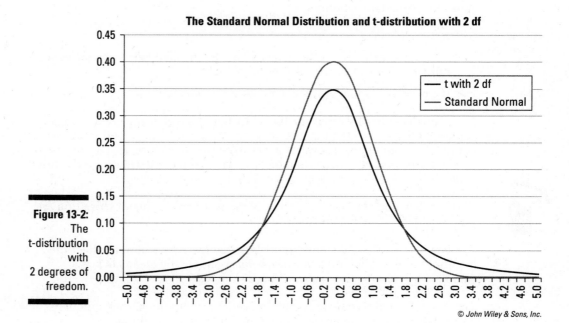

Figure 13-2: The t-distribution with 2 degrees of freedom.

The Standard Normal Distribution and t-distribution with 2 df

© John Wiley & Sons, Inc.

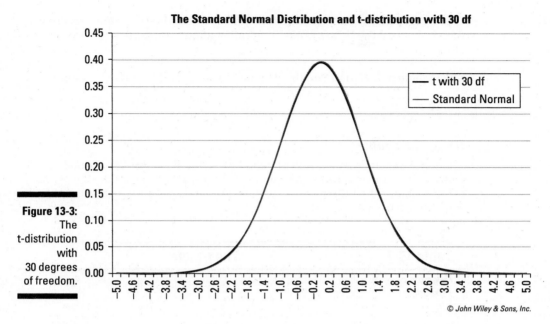

Figure 13-3: The t-distribution with 30 degrees of freedom.

μ represents the mean of a population or a probability distribution. v is the Greek letter *nu*. It is commonly used to designate the number of degrees of freedom of a distribution.

As with the standard normal distribution, the t-distribution is "centered" around 0. The mean isn't affected by the degrees of freedom. The variance of the t-distribution equals the following:

$$\sigma^2 = \frac{v}{v-2}$$

In this formula, v represents the number of degrees of freedom associated with the t-distribution. The formula shows that the variance of the t-distribution is greater than the variance of the standard normal distribution (which is 1). For example, with 10 degrees of freedom, the variance of the t-distribution equals the following:

$$\sigma^2 = \frac{v}{v-2} = \frac{10}{10-2} = \frac{10}{8} = 1.25$$

With 30 degrees of freedom, here's the variance of the t-distribution:

$$\sigma^2 = \frac{v}{v-2} = \frac{30}{30-2} = \frac{30}{28} = 1.07$$

These calculations show that as the number of degrees of freedom increases, the variance of the t-distribution continues to decline, getting progressively closer to 1.

The standard deviation equals the square root of the variance:

$$\sigma = \sqrt{\frac{\nu}{\nu - 2}}$$

Using the t-table

Chapter 4 explains how to calculate the area of tails under the normal distribution. It works the same way here. You can find the value of a given tail area under the t-distribution using a t-table. Table 13-1 is an excerpt taken from the t-table.

Table 13-1		The t-table		
Degrees of Freedom (df)	$t_{0.10}$ $t_{0.20}$	$t_{0.05}$ $t_{0.10}$	$t_{0.025}$ $t_{0.05}$	$t_{0.01}$ $t_{0.02}$	$t_{0.005}$ $t_{0.01}$
8	1.397	1.860	2.306	2.896	3.355
9	1.383	1.833	2.262	2.821	3.250
10	1.372	1.812	2.228	2.764	3.169
11	1.363	1.796	2.201	2.718	3.106
12	1.356	1.782	2.179	2.681	3.055

The top row of the t-table represents tail areas under the t curve. The fact that each column has two headings allows the table to be used for both one- and two-tailed tests. The subscript on the top t in each column reflects the significance level for a one-tailed test. The subscript on the bottom t is the significance level for a two-tailed test.

For example, if you were doing a two-tailed test with the t-distribution at 5% significance, each tail has an area of 2.5%, for a total tail area of 5%. In order to find the appropriate critical values, the relevant column heading is the one that shows $t_{0.025}$ (the area in each tail) and $t_{0.05}$ (the total area in both tails). If the df = 8, then the critical values for the confidence interval would be ± 2.306, which is found in the first row, under the third column. If you were doing a one-tailed test, this same critical value would correspond to a 2.5% significance.

As an alternative to using a t-table, you can find the appropriate values with the Excel TINV function (the inverse of the t-distribution).

The TINV function requires you to supply the total tail area and the degrees of freedom. For the last example, you would use = TINV(0.05, 8) to find the value of 2.306.

The lognormal distribution

The lognormal distribution is derived directly from the normal distribution. Like the normal distribution, the lognormal distribution is uniquely characterized by its mean and standard deviation. Unlike the normal distribution, the lognormal distribution is undefined for negative values.

Applications of the lognormal distribution

The lognormal distribution is used for several important financial applications. For example, you would use it to calculate rates of return to financial assets on a *continuously compounded* basis.

A *continuously compounded* rate of return is compounded an infinite number of times during a given time horizon. A *discretely compounded* rate of return is compounded a finite number of times during a given time horizon.

Daily compounding is an example of discrete compounding. Although rates are normally compounded on a discrete basis in actual practice, continuously compounded rates have several convenient mathematical properties. As a result, they are used in very sophisticated financial models.

Continuously compounded returns are often used in pricing models for *derivative securities,* such as options. These are highly complex products that may be used to manage risk — the models used to price them use extremely advanced math.

The lognormal distribution is positively *skewed* (it has a long right tail) and is undefined for negative values. Figure 13-4 is a graph of the lognormal distribution with a mean of 0 and a standard deviation of 1.

Defining a lognormal random variable

The lognormal distribution is directly related to the normal distribution. As its name implies, a distribution is lognormally distributed if you get a normal distribution when you take its natural logarithm. You can think of this in two ways:

- If $\ln(X)$ is normally distributed, then X is lognormally distributed.
- If X is normally distributed, then e^X is lognormally distributed.

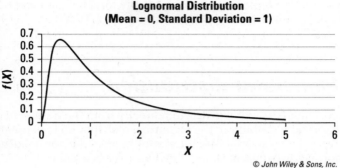

Figure 13-4:
The lognormal distribution with mean = 0 and standard deviation = 1.

For example, you compute the continuously compounded return to an asset as follows:

$$r_t = \ln\left(\frac{P_t}{P_{t-1}}\right)$$

where

ln = The "natural log"

r_t = The continuously compounded return at time t

P_t = The price at time t

P_{t-1} = The price at time $t - 1$, representing one period in the past

P_t / P_{t-1} is known as a *price relative*.

The *natural log* is the logarithm with base *e*. *e* is equal to approximately 2.71828. It represents the exponent to which you would have to raise *e* to get a given value. So for example, $\ln(e) = 1$ and $\ln(e^2) = 2$. The natural logarithm can be calculated for any positive number. It is useful in analyzing compound growth because interest (and returns) generally grows exponentially.

It is often assumed in financial models that rates of return are *normally distributed*. Based on the properties of the lognormal distribution, this implies that the price relatives are *lognormally distributed*.

Moments of the lognormal distribution

The moments of the lognormal distribution are based on the normal distribution from which they are derived.

The mean (expected value) of the lognormal distribution is as follows:

$$\exp(\mu + 0.5\sigma^2)$$

where

μ is the expected value of the corresponding normal distribution.

σ^2 is the variance of the corresponding normal distribution.

exp(x) is an alternative way of expressing e^x; exp often improves legibility in mathematical equations.

You calculate the variance of the lognormal distribution like this:

$$\exp(2\mu + \sigma^2)(\exp(\sigma^2) - 1)$$

The standard deviation of the lognormal distribution is simply the square root of the variance:

$$\sqrt{\exp(2\mu + \sigma^2)(\exp(\sigma^2) - 1)}$$

As an example, suppose Y is a normally distributed random variable with a mean of 1 and a standard deviation of 2. Then $X = e^Y$ is a lognormally distributed random variable. Its mean, variance, and standard deviation are computed as follows:

$$E(X) = \exp(\mu + 0.5\sigma^2) = \exp(1 + 0.5(2)^2) = e^3 = 20.1$$

$$Var(X) = \exp(2\mu + \sigma^2)(\exp(\sigma^2) - 1)$$

$$= \exp(2(1) + (2)^2)(\exp(2)^2 - 1) = \exp(6)\exp(3) = 8,103.08$$

The standard deviation is the square root of the variance of 21,623.037, which is about 147.05.

You can use the Excel function LOGNORM.DIST to find the probability that a lognormally distributed random variable X is less than or equal to a specified value. This is also known as the cumulative probability of X. (The function is LOGNORM.DIST in Excel 2010; in older editions of Excel, the function is LOGNORMDIST.)

The values required for the LOGNORM.DIST function are x (the value whose cumulative probability you are looking for), mean (the mean of the distribution), standard_dev (the standard deviation of the distribution), and cumulative. Cumulative equals 1 if you are solving for a "less than or equal to" probability, and 0 otherwise.

For example, suppose X is lognormal with a mean of 0 and a standard deviation of 1. You compute the probability that X is less than or equal to 1 as LOGNORM.DIST(1, 0, 1, 1) = 0.5.

The chi-square distribution

The chi-square distribution (χ^2) is a *continuous* probability distribution derived from the standard normal distribution. Similar to the t-distribution, the chi-square distribution is uniquely characterized by the number of *degrees of freedom* (df).

The letter χ is the Greek letter *chi* (pronounced "ki").

Applications of the chi-square distribution

You can use the chi-square distribution in several different business applications, such as testing hypotheses about a population variance, conducting goodness of fit tests (to determine whether a population follows a specified distribution), testing for the independence of two populations, and so forth.

The chi-square distribution has one key feature in common with the t-distribution: It's uniquely characterized by its degrees of freedom. Unlike the t-distribution (and the standard normal distribution), the chi-square distribution

✔ Is defined only for positive values.

✔ Is *not* symmetrical about its mean — instead, it is *positively skewed*.

A positively skewed distribution has a long right tail. Figures 13-5 through 13-7 show the chi-square distribution with 5, 10, and 30 df. In each case, the horizontal axis represents different values that may be assumed by a chi-square random variable, and the vertical axis represents the corresponding probabilities.

Each of the graphs in the three figures shows that the distribution is undefined for negative values (notice that there are no negative values along the horizontal axis). Additionally, as the number of degrees of freedom increases, the distribution shifts to the right and begins to resemble the normal distribution. As mentioned earlier, the distribution has a long right tail (it's skewed to the right).

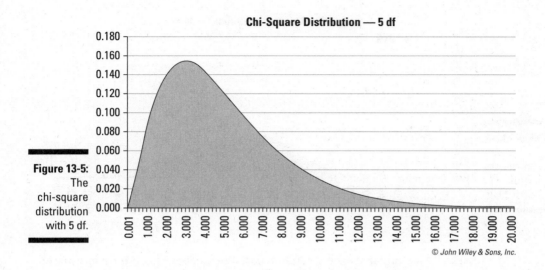

© John Wiley & Sons, Inc.

Figure 13-5: The chi-square distribution with 5 df.

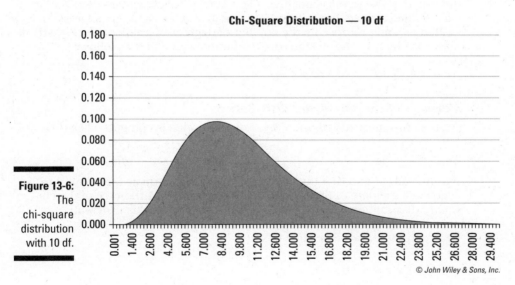

© John Wiley & Sons, Inc.

Figure 13-6: The chi-square distribution with 10 df.

Defining a chi-square random variable

A chi-square random variable is a sum of independent, squared *standard normal* random variables. The following equation expresses this:

$$\chi_v^2 = Z_1^2 + Z_2^2 + Z_3^2 + \ldots + Z_v^2$$

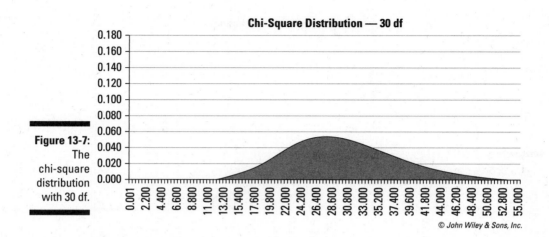

Figure 13-7: The chi-square distribution with 30 df.

Each Z is a standard normal random variable. Because each term is squared, the sum of these terms is positive. (This is why the chi-square distribution isn't defined for negative values.) The number of terms in this expression is written as v and represents the number of degrees of freedom of the distribution. For example, the chi-square distribution with 5 df is as follows:

$$\chi_5^2 = Z_1^2 + Z_2^2 + Z_3^2 + Z_4^2 + Z_5^2$$

Moments of the chi-square distribution

For the chi-square distribution, the expected value is computed like this:

$$E(X) = v$$

This formula shows that the expected value simply equals the number of degrees of freedom of the distribution. For example, in the case of the chi-square distribution with 5 df, the average value under this distribution is 5.

You compute the variance like this:

$$\sigma^2 = 2v$$

This equation shows that the variance equals two times the number of degrees of freedom.

The standard deviation is the *square root* of the variance:

$$\sigma = \sqrt{2v}$$

Using the chi-square table

You find locations in the right tail of the chi-square distribution with a specified tail area with a chi-square table. Table 13-2 shows a portion of the chi-square table.

Table 13-2			The Chi-square Table					
Degrees of Freedom/ Tail Area	0.99	0.975	0.95	0.90	0.10	0.05	0.025	0.01
1	0.000	0.000	0.000	0.016	2.706	3.841	5.024	6.635
2	0.020	0.051	0.103	0.211	4.605	5.991	7.378	9.210
3	0.115	0.216	0.352	0.584	6.251	7.815	9.348	11.345
4	0.297	0.484	0.711	1.064	7.779	9.488	11.143	13.277
5	0.554	0.831	1.145	1.610	9.236	11.070	12.833	15.086

For example, suppose the critical value of a hypothesis test is based on the chi-square distribution with 4 df and a right tail area of 0.05. The appropriate value in the chi-square distribution is 9.488 (found at the intersection of the row labeled 4 and the column labeled 0.05).

As an alternative to using a chi-square table, you can use Excel's built-in function CHISQ.INV (the inverse of the chi-square distribution).

There are two values required for this function. It is invoked as CHISQ.INV (probability, degrees of freedom). One important way in which the table and the Excel function differ is that the table shows areas in the *right* tail of the distribution, whereas the Excel function is based on areas in the *left* tail. As a result, to use the Excel function for areas in the right tail, you must use 1 minus the probability that you are looking for with the Excel function. Based on the last example, CHISQ.INV(1 – 0.05, 4) = CHISQ.INV(0.95, 4) gives a value of 9.488.

The F-distribution

The F-distribution is a *continuous* probability distribution derived from the chi-square distribution. One thing that distinguishes the F-distribution from the chi-square distribution is that the F-distribution has two different types of degrees of freedom, which are known as *numerator degrees of freedom* and *denominator degrees of freedom*. Similar to the chi-square and lognormal distribution, it's positively skewed and is undefined for negative values. Figure 13-8 shows the F-distribution.

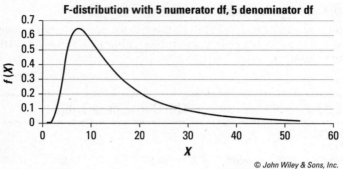

Figure 13-8:
The F distribution with 5 numerator and 5 denominator df.

Applications of the F-distribution

The F-distribution has several different applications in statistical analysis, including the following:

- ✔ Testing hypotheses about the equality of population variances
- ✔ Testing joint hypotheses about the slope coefficients in a regression equation

Defining an F random variable

An F random variable is the ratio of two chi-square random variables, each divided by its own degrees of freedom:

$$F = \frac{\chi_1^2 / v_1}{\chi_2^2 / v_2}$$

where

χ_1^2 and χ_2^2 are chi-square random variables.

v_1 and v_2 are the corresponding degrees of freedom.

v_1 is referred to as the numerator df of the F random variable, and v_2 is referred to as the denominator df.

Moments of the F-distribution

The moments of the F-distribution are computed as follows. Here's the mean (expected value) of the F-distribution:

$$\frac{v_2}{v_2 - 2}, \ v_2 > 2$$

v_2 is the denominator df; the mean is only defined for values of v_2 greater than 2 (to avoid division by 0).

The variance of the F-distribution is calculated like this:

$$\frac{2v_2^2(v_1+v_2-2)}{v_1(v_2-2)^2(v_2-4)}, \ v_1 > 0, \ v_2 > 4$$

The variance is defined only for values of v_1 greater than zero and values of v_2 greater than 4 (to avoid division by 0).

The standard deviation of the F-distribution is simply the square root of the variance:

$$\sqrt{\frac{2v_2^2(v_1+v_2-2)}{v_1(v_2-2)^2(v_2-4)}}, \ v_1 > 0, \ v_2 > 4$$

Using the F-table

You use the F-table to show the location of the right tail of the distribution. To find values of the F-distribution with a table, you must specify both the numerator and denominator degrees of freedom. Therefore, the entire table can be used only for a single area of the right tail.

For example, suppose you're looking for the value under the F-distribution such that the area in the right tail is 5 percent. (This area is often expressed as α, which is the Greek letter *alpha*.) The numerator degrees of freedom are 4, and the denominator degrees of freedom are 3. You can find the appropriate value in Table 13-3 at the intersection of the column headed 4 and the row labeled 3; the value is 9.12.

Table 13-3	A Section of the F-table with $\alpha = 0.05$						
$v_2 \backslash v_1$	2	3	4	5	6	7	8
2	19.00	19.16	19.25	19.30	19.33	19.35	19.37
3	9.55	9.28	9.12	9.01	8.94	8.89	8.85
4	6.94	6.59	6.39	6.26	6.16	6.09	6.04
5	5.79	5.41	5.19	5.05	4.95	4.88	4.82
6	5.14	4.76	4.53	4.39	4.28	4.21	4.15
7	4.74	4.35	4.12	3.97	3.87	3.79	3.73
8	4.46	4.07	3.84	3.69	3.58	3.50	3.44

As an alternative to using the F-table, you can use Excel's built-in function F.INV (the inverse of the F-distribution). The values required for this function are F.INV(probability, deg_freedom1, deg_freedom2).

Unlike the table, the probability in this function refers to the area in the *left* tail. deg_freedom1 represents the numerator degrees of freedom, and deg_freedom2 represents the denominator degrees of freedom.

For example, for the F-distribution with 5 numerator degrees of freedom and 7 denominator degrees of freedom, the location of the 5 percent right tail (which is also the 95 percent left tail), you enter the following: F.INV(1-0.05, 5, 7), which is equal to = F.INV(0.95, 5, 7). The result is 3.97.

That's a lot to digest. But rest assured we use these distributions in ensuing chapters to answer practical questions about financial assets and returns on investment. We also use these distributions to evaluate the accuracy of the predictive models that we discuss.

Chapter 14

To All the Variables We've Encountered: Multivariate Statistical Techniques

*M*ultivariate techniques are used to analyze the relationship between two or more variables. These techniques are of necessity more complex than univariate techniques (which are covered in Chapter 13).

This chapter shows how hypothesis testing can be used to determine the relationship between any and all of the following:

✔ Two population means

✔ Two population variances

✔ Three or more population means

The chapter also discusses correlation in detail, showing two different approaches to computing it:

✔ Pearson's product-moment correlation coefficient

✔ Spearman's rank correlation coefficient

Testing Hypotheses about Two Population Means

Testing a hypothesis about the means of two different populations is conducted in six steps:

1. Stating the null hypothesis

2. Stating the alternative hypothesis

3. Specifying the level of significance

4. Computing the test statistic

5. Determining the critical value or values

6. Making a decision

Before beginning the hypothesis testing procedure, a sample of data is chosen from both populations, and key measures such as the sample mean and sample standard deviation are computed from each. The null hypothesis will be rejected if the sample data is "too extreme" to be consistent with the null hypothesis.

The null hypothesis for two population means

To test the equality of two population means, the null hypothesis is expressed like so:

H_0: $\mu_1 = \mu_2$

The key terms are:

H_0: The null hypothesis

μ_1: The mean of population 1

μ_2: The mean of population 2

The *null hypothesis* is a statement that's assumed to be true unless there is *strong* contrary evidence. In this case, the *alternative hypothesis* is accepted instead.

Alternative hypotheses for two population means

The alternative hypothesis can take one of three forms:

- ✔ Right-tailed test: H_1: $\mu_1 > \mu_2$
- ✔ Left-tailed test: H_1: $\mu_1 < \mu_2$
- ✔ Two-tailed test: H_1: $\mu_1 \neq \mu_2$

The right-tailed test is used if you're looking for strong evidence that the mean of population 1 is greater than the mean of population 2. Equivalently, the left-tailed test is used if you are looking for strong evidence that the mean of population 1 is *less than* the mean of population 2. The two-tailed test is used if you are looking for evidence that the mean of population 1 is either greater than or less than the mean of population 2.

Level of significance

The *level of significance,* usually designated α (Greek *alpha*) is typically chosen to be 0.01, 0.05, or 0.10. This number refers to the likelihood of rejecting the null hypothesis when it is actually true, which is known as a *Type I error* — a *Type II error* occurs when the null hypothesis is false but is not rejected.

The higher the level of significance, the greater the probability of committing a Type I error, but the *lower* the probability of committing a Type II error. The choice of the level of significance depends on the relative importance of avoiding a Type I error relative to avoiding a Type II error. In many business applications, the level of significance is set equal to 0.05.

Test statistics and critical values for testing hypotheses about two population means

The *test statistic* is designed to show whether the sample data that has been chosen is consistent with the null hypothesis. When you're testing hypotheses about two population means, the appropriate test statistic depends on whether the following are true:

- ✔ The samples drawn from the two populations are *independent* of each other.
- ✔ The variances of the two populations are equal.
- ✔ The samples chosen from the two populations are large or small.

For this type of hypothesis test, a *small* sample contains fewer than 30 observations; a *large* sample contains at least 30 observations.

The dividing line between small and large samples comes from the Central Limit Theorem, which shows that the distribution of the means of large samples is approximately normal.

Samples are *independent* if the selection of the first sample has no bearing on the selection of the second. For example, suppose you want to compare gas prices in New York and New Jersey. You might select a random sample of gas stations from each state. Because the stations are in different states, the selection of the New Jersey gas stations has no effect on which stations are selected from New York. Thus the samples are independent.

On the other hand, if volunteers are trying out a new drug to determine whether it can reduce cholesterol, the readings of the volunteers prior to taking the drug and after taking the drug could be thought of as two separate *dependent* populations. In fact, once the members of the first sample are chosen, the members of the second sample are completely determined.

For independent populations, there are three relevant scenarios:

- ✔ The variances of the two populations are equal (or assumed to be so).
- ✔ The variances of the two populations are unequal, and at least one sample is small.
- ✔ The variances of the two populations are unequal, and both samples are large.

Independent populations

When using samples from independent populations, if the variances of the two populations are equal, the test statistic is automatically based on the Student's t-distribution. The sizes of the samples drawn from the two populations are irrelevant.

Equal population variances

If the variances of two populations are equal (or are assumed to be equal), the appropriate test statistic is based on the Student's t-distribution:

$$t = \frac{(\bar{x}_1 - \bar{x}_2) - (\mu_1 - \mu_2)_0}{\sqrt{s_p^2 \left(\frac{1}{n_1} + \frac{1}{n_2} \right)}}$$

The key terms in this formula are as follows:

$\bar{x}_1$ is the mean of the sample chosen from population 1.

$\bar{x}_2$ is the mean of the sample chosen from population 2.

μ_1 is the mean of population 1.

μ_2 is the mean of population 2.

$(\mu_1 - \mu_2)_0$ is the hypothesized difference between populations 1 and 2; this equals 0 when the population means are hypothesized to be equal.

n_1 is the size of the sample chosen from population 1.

n_2 is the size of the sample chosen from population 2.

$s^2{}_1$ is the variance of the sample chosen from population 1.

$s^2{}_2$ is the variance of the sample chosen from population 2.

$s^2{}_p$ is the estimated "pooled" variance of the two populations; this is the average variance of the two populations and is computed as follows:

$$s_p^2 = \frac{(n_1 - 1)s_1^2 + (n_2 - 1)s_2^2}{n_1 + n_2 - 2}$$

The critical values are:

- ✔ Two-tailed test: $\pm t_{\alpha/2}^{n_1 + n_2 - 2}$
- ✔ Right-tailed test: $t_\alpha^{n_1 + n_2 - 2}$
- ✔ Left-tailed test: $-t_\alpha^{n_1 + n_2 - 2}$

The degrees of freedom in each case are: $n_1 + n_2 - 2$.

Unequal population variances: At least one sample is small

If the variances of two populations are *not* equal and at least one sample is small (less than 30), the appropriate test statistic is as follows:

$$t = \frac{(\bar{x}_1 - \bar{x}_2) - (\mu_1 - \mu_2)_0}{\sqrt{\left(\dfrac{s_1^2}{n_1} + \dfrac{s_2^2}{n_2}\right)}}$$

In this case, the critical values are drawn from the t-distribution with degrees of freedom equal to

$$df = \frac{[(s_1^2 / n_1) + (s_2^2 / n_2)]^2}{\dfrac{(s_1^2 / n_1)^2}{(n_1 - 1)} + \dfrac{(s_2^2 / n_2)^2}{(n_2 - 1)}}$$

This number must be rounded to the nearest integer.

Unequal population variances: Both sample sizes are large

If the variances of two populations are *not* equal, and both samples are large (30 or greater), the appropriate test statistic is based on the *standard normal* distribution:

$$Z = \frac{(\bar{x}_1 - \bar{x}_2) - (\mu_1 - \mu_2)_0}{\sqrt{\left(\dfrac{s_1^2}{n_1} + \dfrac{s_2^2}{n_2}\right)}}$$

There are an infinite number of normal distributions, each with its own unique mean and standard deviation. The *standard* normal distribution has a mean of 0 and a standard deviation of 1.

The critical values are as follows:

- Two-tailed test: $\pm Z_{\alpha/2}$
- Right-tailed test: Z_α
- Left-tailed test: $-Z_\alpha$

The decision

The decision as to whether or not a null hypothesis should be rejected is determined as follows:

- For a *right-tailed test,* the null hypothesis is rejected if the test statistic *is greater than* the critical value.

- For a *left-tailed test,* the null hypothesis is rejected if the test statistic *is less than* the critical value.

- For a *two-tailed test,* the null hypothesis is rejected if the test statistic is *greater than* the positive critical value or *less than* the negative critical value.

The case of dependent populations

If samples are chosen from two *dependent* populations, they are referred to as *paired* samples. In this case, the null hypothesis is based on the *differences* between the sample elements. (You can think of this as a test of a single population consisting of the differences between the original populations.) In this case, the notation is slightly different. For example, the null hypothesis is written like this:

$$H_0: \mu_d = 0$$

where:

μ_d represents the mean difference between the two populations;
$\mu_d = \mu_1 - \mu_2$.

The three possible alternative hypotheses are:

✔ **Right-tailed test:** $H_1: \mu_d > 0$. In this case, the alternative hypothesis is that the mean of population 1 is *greater than* the mean of population 2.

✔ **Left-tailed test:** $H_1: \mu_d < 0$. In this case, the alternative hypothesis is that the mean of population 1 is *less than* the mean of population 2.

✔ **Two-tailed test:** $H_1: \mu_d \neq 0$. In this case, the alternative hypothesis is that the means of populations 1 and 2 are *not equal*.

For paired samples, the test statistic is always based on the Student's t-distribution:

$$t = \frac{\bar{d}}{s_d / \sqrt{n}}$$

The key terms in this equation are:

✔ $\bar{d}$ = the average (mean) difference between paired samples

✔ s_d = the standard deviation of the sample differences

You compute the mean difference in the same way as the sample mean when the sample is drawn from a single population:

$$\bar{d} = \frac{\sum_{i=1}^{n} d_i}{n}$$

This formula indicates that you calculate the average difference of the paired samples by adding up all the individual differences and then dividing by the total number of samples.

The standard deviation of the sample differences is computed like this:

$$s_d = \sqrt{\frac{\sum_{i=1}^{n} d_i^2 - n\bar{d}^2}{n-1}}$$

With paired samples, the critical values are as follows:

- ✓ Two-tailed test: $\pm t_{\alpha/2}^{n-1}$
- ✓ Right-tailed test: t_{α}^{n-1}
- ✓ Left-tailed test: $-t_{\alpha}^{n-1}$

The degrees of freedom in each case is $n-1$; n is the number of pairs of observations chosen from the two populations.

As an example, suppose that Twinings Tea decides to increase tea consumption in the United States by embarking on a major advertising campaign. In order to determine the effectiveness of the campaign, a sample of six consumers is chosen. Table 14-1 shows the consumers' annual spending on tea before and after campaign advertising.

Table 14-1	Paired Differences between Two Samples			
Consumer	*Spending Prior to Advertisements (x_1)*	*Spending After Advertisements (x_2)*	$d_i = x_1 - x_2$	d_i^2
1	150	160	−10	100
2	225	240	−15	225
3	230	225	5	25
4	180	195	−15	225
5	160	165	−5	25
6	90	95	−5	25
Sum	**1,035**	**1,080**	**−45**	**625**

The column labeled $d_i = x_1 - x_2$ shows the differences between spending prior to the advertisements and spending after for each consumer. (A negative difference indicates that the consumer is spending more after the advertising takes place.) The column labeled d_i^2 shows the squared differences.

The company tests the null hypothesis that tea consumption remains unchanged after the advertising campaign takes place at the 5 percent level of significance as follows:

The null hypothesis is:

$$H_0: \mu_1 = \mu_2$$

Because Twinings is looking for strong evidence that the spending per consumer *increased* after the advertising campaign, this is a *left-tailed* test. (In other words, the company hopes to find that spending has increased, which implies a negative difference.)

The alternative hypothesis is this:

$H_1: \mu_1 < \mu_2$

You would compute the test statistic by substituting the values in the table as follows (note that $n = 6$ because there are six paired samples):

$$\bar{d} = \frac{-45}{6} = -7.50$$

$$s_d = \sqrt{\frac{\sum_{i=1}^{n} d_i^2 - n\bar{d}^2}{n-1}} = \sqrt{\frac{625 - (6)(7.50)^2}{6-1}} = 7.58$$

$$t = \frac{\bar{d}}{s_d / \sqrt{n}} = \frac{7.50}{7.50 / \sqrt{6}} = -2.42$$

Based on the table for the Student's t-distribution, the critical value is:

$$-t_\alpha^{n-1} = -t_{0.05}^6 = -1.943$$

Because the test statistic (–2.42) falls below the critical value (–1.943), the null hypothesis is *rejected* in favor of the alternative hypothesis. That is, the difference between the consumption of tea prior to the advertising campaign and after the campaign is *negative,* which indicates that the advertising campaign has been successful.

Using Analysis of Variance (ANOVA) to Test Hypotheses about Population Means

Analysis of variance (acronymed as ANOVA) is a technique you can use to test hypotheses about the equality of two or more population means. The ANOVA process is based on observations taken from *independent* subjects. For example, a corporation could use ANOVA to compare the mean sales of its divisions in the United States, Canada, and Mexico.

As an example, suppose a researcher wants to compare the grade point averages (GPAs) of students who are attending business schools, schools of arts and science, and performing arts schools. He suspects that GPAs are highest at business schools and wants to determine whether this is true or not.

A sample of GPAs is chosen at four universities; the mean values are shown in Table 14-2.

Table 14-2	Grade Point Average, by School		
	Business Administration	Arts and Sciences	Performing Arts
University 1	3.10	2.90	3.40
University 2	2.90	3.00	3.30
University 3	3.30	2.70	2.90
University 4	3.00	3.10	3.00

In this case, the GPA is considered to be the dependent variable. The three different types of schools are referred to as *treatments*. In this case, differences in GPAs may be attributed to the following:

✔ Variation *between* groups (that is, between schools)

✔ Variation *within* groups (variation among students in the same school)

Here is the null hypothesis:

$$H_0 : \mu_1 = \mu_2 = \mu_3$$

The alternative hypothesis is that the means are *not* all equal.

In this case, assume that the level of significance is set equal to 0.05 (that is, 5 percent).

The test statistic for the ANOVA process is quite complex to calculate. Within the formulas that will be used to compute the test statistic, each element in the table will be designated X_{ij}. The subscripts i and j represent the row and column, respectively, in which each value is found. The subscripts for the data is shown in Table 14-3.

For example, X_{32} is the value found at the intersection of the third row and second column of the table, whereas X_{12} is found at the intersection of the first row and second column.

Table 14-3	Grade Point Averages Shown with Subscripts		
	Business Administration	Arts and Sciences	Performing Arts
University 1	X_{11}	X_{12}	X_{13}
University 2	X_{21}	X_{22}	X_{23}
University 3	X_{31}	X_{32}	X_{33}
University 4	X_{41}	X_{42}	X_{43}

Computing the sum of squared errors (SSE)

In order to construct the test statistic, the first step is to compute the means for each column. You compute the mean of column 1 as follows:

$$\bar{X}_{\bullet 1} = \frac{\sum_{i=1}^{n} X_{i1}}{n_1}$$

where:

$\bar{X}_{\bullet 1}$ = The mean of column 1 (the bar indicates that this is a mean); the subscripts indicate that this average is computed from all elements within column 1.

X_{i1} = The value of X in row i and column 1.

n_1 = The number of elements in column 1.

In this example,

$$\bar{X}_{\bullet 1} = \frac{\sum_{i=1}^{n} X_{i1}}{n_1} = \frac{3.10 + 2.90 + 3.30 + 3.00}{4} = \frac{12.30}{4} = 3.075$$

The means of columns 2 and 3 are computed as follows:

$$\bar{X}_{\bullet 2} = \frac{\sum_{i=1}^{n} X_{i2}}{n_2} = \frac{2.90 + 3.00 + 2.70 + 3.10}{4} = \frac{11.70}{4} = 2.925$$

$$\bar{X}_{\bullet 3} = \frac{\sum_{i=1}^{n} X_{i3}}{n_3} = \frac{3.40 + 3.30 + 2.90 + 3.00}{4} = \frac{12.60}{4} = 3.150$$

The next step is to subtract the mean of each column from each element within that column and then square the result. The calculations are shown in Table 14-4.

Table 14-4	Grade Point Averages: Differences from the Column Means		
	Business Administration	Arts and Sciences	Performing Arts
University 1	$(3.10 - 3.075)^2$ $= 0.000625$	$(2.90 - 2.925)^2$ $= 0.000625$	$(3.40 - 3.150)^2$ $= 0.062500$
University 2	$(2.90 - 3.075)^2$ $= 0.030625$	$(3.00 - 2.925)^2$ $= 0.005625$	$(3.30 - 3.150)^2$ $= 0.022500$
University 3	$(3.30 - 3.075)^2$ $= 0.050625$	$(2.70 - 2.925)^2$ $= 0.050625$	$(2.90 - 3.150)^2$ $= 0.062500$
University 4	$(3.00 - 3.075)^2$ $= 0.005625$	$(3.10 - 2.925)^2$ $= 0.030625$	$(3.00 - 3.150)^2$ $= 0.022500$
Sum	0.0875	0.0875	0.1700

The sum of these column totals is $0.0875 + 0.0875 + 0.1700 = 0.3450$. This sum is known as the *error sum of squares (SSE)*.

Computing the sum of squares for treatment (SSTR)

The next step is to compute the *treatment sum of squares (SSTR)*. This requires the calculation of the overall average for the sample, known as the *overall mean* or *grand mean*.

In this example, there are 12 total observations (four universities and three schools). The sum of the 12 observations is as follows:

$$3.10 + 2.90 + 3.30 + 3.00 + 2.90 + 3.00 + 2.70 + 3.10 + 3.40 + 3.30 + 2.90 + 3.00 = 36.6$$

Dividing this by 12 gives the overall mean of $36.6 \div 12 = 3.05$.

You compute the treatment sum of squares (SSTR) with the following steps:

For each column:

- ✔ Compute the squared difference between the column mean and the overall mean.
- ✔ Multiply the result by the number of elements in the column.

SSTR is the sum of these calculations.

In this case, SSTR equals

$$SSTR = 4(3.075 - 3.05)^2 + 4(2.925 - 3.05)^2 + 4(3.15 - 3.05)^2 = 0.105$$

Computing the total sum of squares (SST)

The *total sum of squares (SST)* equals the sum of the treatment sum of squares (SSTR) and the error sum of squares (SSE). This is shown as follows:

$$SST = SSTR + SSE = 0.105 + 0.345 = 0.45$$

Computing the mean sums of squares

The mean sums of squares are the average values of the corresponding sums of squares. The two mean sums of squares that are needed for this hypothesis test are as follows:

- ✔ Treatment mean square (MSTR)
- ✔ Error mean square (MSE)

MSTR is computed according to the following formula:

$$MSTR = \frac{SSTR}{t-1}$$

In this formula, t is the number of treatments (in this case, the number of types of schools). MSTR is the mean of the treatment sum of squares (SSTR).

In this example,

$$MSTR = \frac{SSTR}{t-1} = \frac{0.105}{3-1} = 0.0525$$

MSE is computed according to the following formula:

$$MSE = \frac{SSE}{N-t}$$

In this formula, N is the total number of observations in the sample. *MSE* is the mean of the error sum of squares (SSE). In this example,

$$MSE = \frac{SSE}{N-t} = \frac{0.345}{12-3} = 0.0383$$

Computing the F-statistic

The test statistic for the ANOVA process follows the F-distribution; it is therefore referred to as an *F-statistic.* This is computed as the ratio of MSTR to MSE, as follows:

$$F = \frac{MSTR}{MSE} = \frac{0.0525}{0.0383} = 1.37$$

This value is compared to the critical value to determine if the null hypothesis should be rejected.

The F-Distribution

The F-distribution is a continuous probability distribution, named after the statistician Sir Ronald Fisher (1890–1962). Chapter 13 introduces the F-distribution and some of its properties.

Two of the most important properties of the F-distribution are the following:

✔ It is defined only for positive values.

✔ It is *positively skewed.*

The F-distribution is characterized by two numerical values, or *parameters,* which are known as *degrees of freedom.* There are two types of degrees of freedom that characterize the F-distribution:

✔ Numerator degrees of freedom

✔ Denominator degrees of freedom

Figure 14-1 shows a graph of the F-distribution for different combinations of numerator and denominator degrees of freedom. In each case, numerator degrees of freedom are listed first, and denominator degrees of freedom are listed second (for example, 1,5). The level of significance in each case is 0.05.

The graph shows that the distribution is undefined for negative values. Additionally, as the number of degrees of freedom increases, the *shape* of the distribution changes; it shifts to the right and begins to resemble the normal distribution.

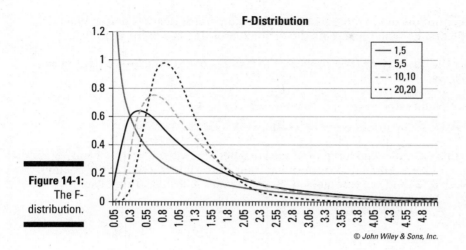

Figure 14-1:
The F-
distribution.

© John Wiley & Sons, Inc.

Finding the critical values using the F-table

Because the F-distribution is based on two different types of degrees of free-dom, there is one table for each possible value of α (the level of significance). Table 14-5 shows the different values of the F-distribution corresponding to a 0.05 (5 percent) level of significance.

Table 14-5	The F-distribution with $\alpha = 0.05$							
$\nu_2 \backslash \nu_1$	2	3	4	5	6	7	8	9	10
2	19.00	19.16	19.25	19.30	19.33	19.35	19.37	19.38	19.40
3	9.55	9.28	9.12	9.01	8.94	8.89	8.85	8.81	8.79
4	6.94	6.59	6.39	6.26	6.16	6.09	6.04	6.00	5.96
5	5.79	5.41	5.19	5.05	4.95	4.88	4.82	4.77	4.74
6	5.14	4.76	4.53	4.39	4.28	4.21	4.15	4.10	4.06
7	4.74	4.35	4.12	3.97	3.87	3.79	3.73	3.68	3.64
8	4.46	4.07	3.84	3.69	3.58	3.50	3.44	3.39	3.35
9	4.26	3.86	3.63	3.48	3.37	3.29	3.23	3.18	3.14
10	4.10	3.71	3.48	3.33	3.22	3.14	3.07	3.02	2.98

The top row of the table represents the numerator degrees of freedom (v_1). The first column represents the denominator degrees of freedom (v_2).

For the one-way ANOVA process, the degrees of freedom are computed as follows:

- Numerator degrees of freedom $= t - 1 = 3 - 1 = 2$
- Denominator degrees of freedom $= N - t = 12 - 3 = 9$

In this example, find a right-tail area of 5 percent with $v_1 = t - 1 = 2$ and $v_2 = N - t = 9$. You'll find this critical value at the intersection of the column labeled "2" and the row labeled "9" — it equals 4.26. It's written like this:

$$F_\alpha^{v_1, v_2} = F_{0.05}^{2,9} = 4.26$$

Making a decision

The one-way ANOVA hypothesis test is a right-tailed test; if the test statistic exceeds the critical value, the null hypothesis that all population means are equal is *rejected;* otherwise, it is not.

In this example, the test statistic equals 1.37, whereas the critical value equals 4.26. Because the test statistic does *not* exceed the critical value, the null hypothesis that the three population means are equal is *not* rejected.

This shows that the means of the GPAs among students in the schools of business administration, schools of arts and science, and performing arts schools are *equal.*

F-Test for the Equality of Two Population Variances

Hypothesis testing for the variance of two populations is based on the F-distribution. The appropriate steps are similar to other hypothesis tests. The main differences are in the form of the null and alternative hypotheses, as well as the specific test statistic and critical values that are used.

Null hypothesis

The null hypothesis is written as follows:

$$H_0: \sigma_1^2 = \sigma_2^2$$

where:

σ_1^2 = The variance of population 1

σ_2^2 = The variance of population 2

The null hypothesis shows that the population variances are assumed to be equal unless there is strong contradictory evidence against this assumption.

Alternative hypothesis

The three possible alternative hypotheses are as follows:

For a right-tailed test: $H_1: \sigma_1^2 > \sigma_2^2$

For a left-tailed test: $H_1: \sigma_1^2 < \sigma_2^2$

For a two-tailed test: $H_1: \sigma_1^2 \neq \sigma_2^2$

The *right-tailed* test shows whether there's sufficient evidence to conclude that the variance of population 1 is *greater than* the variance of population 2; the *left-tailed* test shows whether there's sufficient evidence to conclude that the variance of population 1 is *less than* the variance of population 2. The *two-tailed test* shows whether or not the population variances are equal.

Level of significance

The level of significance (α) must be specified in order to conduct a hypothesis test; this equals the probability of a Type I error.

Test statistic

The test statistic equals the ratio of the sample variances (the larger of the two is always put in the numerator):

$$F = \frac{s_1^2}{s_2^2}$$

where

$$s_1^2 = \text{Sample variance 1}$$
$$s_2^2 = \text{Sample variance 2}$$
$$s_1^2 > s_2^2$$

Critical values

For this type of hypothesis test, the critical value is taken from the F-distribution. The critical values of the test correspond to the alternative hypothesis that is chosen, as follows:

Right-tailed test: Critical value $= F_\alpha^{v_1, v_2}$

Left-tailed test: Critical value $= F_\alpha^{v_1, v_2}$

Two-tailed test: Critical value $= F_{\alpha/2}^{v_1, v_2}$

where

$v_1 =$ Numerator degrees of freedom; this equals $n_1 - 1$, where n_1 is the size of the sample drawn from population 1.

$v_2 =$ Denominator degrees of freedom; this equals $n_2 - 1$, where n_2 is the size of the sample drawn from population 2.

Decision rule

The decision rule depends on the type of alternative hypothesis that is being used. In each case, if the test statistic is more "extreme" than the critical value, the null hypothesis will be rejected.

The decision rule can be summarized as follows:

✔ For a right-tailed test, the null hypothesis will be rejected if $F > F_\alpha^{v_1, v_2}$.

✔ For a left-tailed test, the null hypothesis will be rejected if $F > F_\alpha^{v_1, v_2}$.

✔ For a two-tailed test, the null hypothesis will be rejected if $F > F_{\alpha/2}^{v_1, v_2}$.

As an example, suppose an investor wants to determine whether two portfolios have the same volatility. A sample of ten stocks is taken from both portfolios; the sample standard deviation of Portfolio 1 is 0.26 (26 percent), and the sample standard deviation of Portfolio 2 is 0.24 (24 percent). The null and alternative hypotheses are like this:

$$H_0: \sigma_1^{\,2} = \sigma_2^{\,2}$$
$$H_1: \sigma_1^{\,2} \neq \sigma_2^{\,2}$$

Assume that the level of significance is 10 percent ($\alpha = 0.10$). The test statistic is as follows:

$$F = \frac{s_1^2}{s_2^2}$$

Substituting the appropriate values gives the following result:

$$F = \frac{s_1^2}{s_2^2} = \frac{(0.26)^2}{(0.24)^2} = 1.174$$

Because this is a two-tailed test with a 10 percent level of significance, with two samples of ten each, you compute the numerator and denominator degrees like so:

$$v_1 = n_1 - 1 = 10 - 1 = 9$$
$$v_2 = n_2 - 1 = 10 - 1 = 9$$

Table 14-5 contains the appropriate critical value (3.18):

$$F_{\alpha/2}^{v_1, v_2} = F_{0.05}^{9,9} = 3.18$$

Because the test statistic equals $F = 1.174$, and the critical value is 3.18, this shows that

$$F < F_{\alpha/2}^{v_1, v_2}$$

Therefore, the null hypothesis fails to be rejected. There is insufficient evidence to reject the proposition that the volatilities of the two portfolios are equal.

Correlation

Correlation is a measure of the *dependency* between two variables. Variables that tend to rise or fall at the same time are said to be *positively* correlated; variables that tend to move in opposite directions are said to be *negatively* correlated; and variables that are not related to each other are said to be *uncorrelated* or *independent*.

For example, the number of runs scored by a baseball team during a season and the number of games that it wins are usually positively correlated. The number of runs allowed by a team's pitching staff and the number of games that the team wins are usually negatively correlated. Most likely, the mean daily temperature during the season and the number of games won by the team would be uncorrelated.

You can compute correlation in several different ways. Here are two of the more commonly used techniques:

- Pearson's product-moment correlation coefficient
- Spearman's rank correlation coefficient

Pearson's product-moment correlation coefficient

The Pearson's product-moment correlation coefficient is computed for two samples, X and Y, as follows:

$$r_{XY} = \frac{\sum_{i=1}^{n}(X_i - \bar{X})(Y_i - \bar{Y})}{\sqrt{\sum_{i=1}^{n}(X_i - \bar{X})^2}\sqrt{\sum_{i=1}^{n}(Y_i - \bar{Y})^2}}$$

where

r_{XY} = Pearson's product-moment correlation coefficient

n = Number of elements in each sample

X_i = An element in sample X

Y_i = An element in sample Y

$\bar{X}$ = Mean of sample X

$\bar{Y}$ = Mean of sample Y

Based on this formula, this correlation coefficient can only assume values between –1 and 1. A value of –1 indicates that X and Y are *perfectly negatively correlated*, and a value of 1 indicates that X and Y are *perfectly positively correlated*.

As an example, Table 14-6 shows the prices of Stock X and Stock Y during five years.

Table 14-6	Stock Prices 2010–2014	
Year	**Stock X**	**Stock Y**
2010	100	50
2011	98	48
2012	114	56
2013	108	61
2014	105	65

The mean price of Stock X:

$$\bar{X} = \frac{\sum_{i=1}^{n} X_i}{n} = \frac{100 + 98 + 114 + 108 + 105}{5} = \frac{525}{5} = 105$$

The mean price of Stock Y:

$$\bar{Y} = \frac{\sum_{i=1}^{n} Y_i}{n} = \frac{50 + 48 + 56 + 61 + 65}{5} = \frac{280}{5} = 56$$

The remaining calculations are shown in Table 14-7.

Table 14-7		Calculations for Pearson's Product-Moment Correlation Coefficient			
Year	**Stock X (Xi)**	**Stock Y (Yi)**	$\left(X_i - \bar{X}\right)^2$	$\left(Y_i - \bar{Y}\right)^2$	$\left(X_i - \bar{X}\right)\left(Y_i - \bar{Y}\right)$
2010	100	50	25	36	30
2011	98	48	49	64	56
2012	114	56	81	0	0
2013	108	61	9	25	15
2014	105	65	0	81	0
Sum	**525**	**280**	**164**	**206**	**101**

The numbers are substituted into the formula as follows:

$$r_{XY} = \frac{\sum_{i=1}^{n}(X_i - \bar{X})(Y_i - \bar{Y})}{\sqrt{\sum_{i=1}^{n}(X_i - \bar{X})^2} \sqrt{\sum_{i=1}^{n}(Y_i - \bar{Y})^2}} = \frac{101}{(\sqrt{264})(\sqrt{206})} = 0.55$$

This shows that the prices of Stocks X and Y are positively correlated. Both stocks tend to do well during the same years and poorly during the same years.

You can use the same formula to compute the correlation between two populations; in that case, the correlation is written as ρ_{XY}.

ρ is the Greek letter *rho*.

Spearman's rank correlation coefficient

Spearman's rank correlation coefficient (also known as Spearman's rho) is a *non-parametric* measure of dependency between two variables.

A *non-parametric measure* isn't based on any distributional assumptions. Pearson's product-moment correlation coefficient is valid only for certain probability distributions, including the normal and the Student's t-distribution.

You compute the Spearman's rank correlation coefficient by converting sample data into *ranks*. For example, suppose that a sample of six supermarkets in New York and six supermarkets in New Jersey are randomly chosen, and the price of a pound of the store's brand of coffee is recorded. X represents the prices at the New York supermarkets, and Y represents the prices at the New Jersey supermarkets. Table 14-8 shows the individual values.

Table 14-8	Sample Coffee Prices		
New York Supermarket	*Price of a Pound of Coffee (X_i)*	*New Jersey Supermarket*	*Price of a Pound of Coffee (Y_i)*
1	$2.99	1	$3.89
2	$3.29	2	$2.59
3	$2.79	3	$2.89
4	$3.49	4	$3.39
5	$2.89	5	$3.49
6	$3.09	6	$3.59

The sorted values of X_i are as follows:

   2.79, 2.89, 2.99, 3.09, 3.29, 3.49

The sorted values of Y_i are as follows:

   2.59, 2.89, 3.39, 3.49, 3.59, 3.89

These values are substituted into the following equation, which defines the Spearman rank correlation coefficient:

$$\rho = 1 - \frac{6 \sum_{i=1}^{n} d_i^2}{n(n^2 - 1)}$$

where

   ρ = Spearman's rank correlation coefficient

   d_i = Difference between the ranks for observation i; that is, $X_i - Y_i$

   n = Number of elements in each sample

Table 14-9 shows the calculations. The values of X and Y have been sorted.

Table 14-9		Calculations for Spearman's Rank Correlation Coefficient			
New York Super-market Sorted Ranks	Price of a Pound of Coffee (X_i)	New Jersey Super-market Sorted Ranks	Price of a Pound of Coffee (Y_i)	Rank Difference $d_i = X_i - Y_i$	Squared Rank Difference d_i^2
3	2.79	2	2.59	1	1
5	2.89	3	2.89	2	4
1	2.99	4	3.39	−3	9
6	3.09	5	3.49	1	1
2	3.29	6	3.59	−4	16
4	3.49	1	3.89	3	9
Sum				0	40

The values from the last column are then substituted into the equation for the Spearman's rank correlation coefficient:

$$\rho = 1 - \frac{6\sum_{i=1}^{n} d_i^2}{n(n^2 - 1)} = 1 - \frac{6(40)}{6(6^2 - 1)} = 1 - \frac{240}{210} = -0.143$$

This shows that there's a negative correlation between the coffee prices at the two supermarket chains.

That brings us to the end of another technical chapter. If you've gotten this far, your head may be spinning a bit. But be of good cheer. In actual business applications, you won't have to make these calculations by hand. Statistical software will do the heavy lifting for you. We've found, however, that having some sense of what's going on inside the black box of your statistical package will make the results more meaningful to you.

Chapter 15

Regression Analysis

*R*egression analysis is a statistical framework that is used to estimate the strength and direction of the relationship between two or more variables. Simple regression analysis is used to estimate the relationship between a dependent variable (Y) and an independent variable (X). Multiple regression analysis is used to estimate the relationship between a dependent variable and *two or more* independent variables. We typically think of the independent variable as something you are trying to predict and the dependent variables as quantities you can measure. Regression analysis is heavily used in economics and finance to understand relationships between variables such as interest rates, GDP growth rates, stock prices, corporate profits, and more.

This chapter covers the methodology that is used to implement and interpret the results of regression analysis. It explains the statistical assumptions underlying regression analysis in detail, and includes a consideration of the problems that may arise with regression analysis. It also introduces two more advanced regression techniques: stepwise regression and logistic regression.

You use *stepwise regression* to find the model that best fits or explains a given dataset. You use *logistic regression* (sometimes known as *logit* regression) when the dependent variable can only assume a small number of possible values. This chapter introduces *binary* logistic regression, where the dependent variable can only assume one of two values — male or female, for example.

The Fundamental Assumption: Variables Have a Linear Relationship

Regression analysis is based on the assumption that there's a *linear relationship* between two random variables X and Y. If this assumption is false, an alternate technique must be used; otherwise, the results will be highly unreliable.

If the relationship between two variables can be expressed in the following form, they are said to be *linearly related:*

$$Y = mX + b$$

Y = The dependent variable

X = The independent variable

m = The *slope* (this equals the change in Y divided by the change in X)

b = The *intercept* (this is the value of Y when X equals 0)

To determine whether a relationship is linear, one quick and easy test is to construct a *scatter plot* of the data. This makes it easy to see the relationship between the two variables at a glance. Figure 15-1, a little later in the chapter, shows what a scatter plot looks like. In the case of a linear relationship, the scatter plot will show the data points lying roughly along a straight line.

Defining the Population Regression Equation

The *population regression equation* or *population regression line* is an equation that best "fits" or "explains" the relationship between X and Y in a population. You write the equation like this:

$$Y_i = \beta_0 + \beta_1 X_i + \varepsilon_i$$

β_0 and β_1 are known as *coefficients* of the regression line. β_1 is the *slope* coefficient, and β_0 is the *intercept* coefficient (or simply the *intercept*). The slope coefficient is by far the more important of the two; it shows how sensitive Y is to changes in X.

The other terms in the equation are as follows:

i = An index; this is used to identify the individual members of the population.

Y_i = A single value of Y, indexed by i, in a population of size n, with the values of Y expressed as $Y_1, Y_2, Y_3, \ldots, Y_n$

X_i = A single value of X, indexed by i, in a population of size n, with the values of X expressed as $X_1, X_2, X_3, \ldots, X_n$

ε_i = An "error term," indexed by i, as each observation in the population (X_i, Y_i) has an error term associated with it

In the population regression equation, there's a slope and an intercept, but there's also one additional term that you don't normally find in the equation for a straight line — the *error term*. The error term is included because the population regression equation doesn't perfectly capture the relationship between X and Y. (This is because factors other than X may account for the value of Y.) For each (X_i, Y_i) pair, the error term ε_i equals the difference between the actual value of Y_i and the value predicted by the regression line. The error terms may be positive or negative; on average, they equal zero.

Estimating the Population Regression Equation

In most situations, estimating the population regression line with data from the entire population would be impractical, because it's usually a very time-consuming and expensive process. Instead, you can draw a *sample* from the underlying population.

It's extremely important to ensure that any samples that are drawn truly reflect the properties of the underlying population! Otherwise, the results are unlikely to be accurate.

You use the sample data to construct a *sample regression equation* or *sample regression line,* which is an *estimate* of the population regression equation. The sample regression equation is expressed like this:

$$\hat{Y}_i = \hat{\beta}_0 + \hat{\beta}_1 X_i$$

The terms in this equation are as follows:

$\hat{Y}_i$, the estimated value of Y_i

$\hat{\beta}_0$, the estimated value of β_0

$\hat{\beta}_1$, the estimated value of β_1

TIP

The symbol $^\wedge$ is often used in statistics to indicate an *estimated value*. The proper name of this punctuation mark is *caret*. Informally, it's called a "hat." For example, you pronounce $\hat{\beta}_0$ as "beta zero hat."

You determine the estimated values for $\hat{\beta}_0$ and $\hat{\beta}_1$ by minimizing the sum of the *squared* differences between the observed Y values in the sample and those predicted by the sample regression equation, as shown in the following equation:

$$\min \sum_{i=1}^{n} \left(Y_i - \hat{Y}_i \right)^2$$

In this formula, *min* stands for *minimize*.

The difference between the actual value of Y_i and the predicted value of Y_i is known as a *residual*:

$$\hat{\varepsilon}_i = Y_i - \hat{Y}_i$$

You use $\hat{\varepsilon}_i$ to represent the residual associated with a single observation (X_i, Y_i). Think of the residual as an estimated error term.

Minimizing the residuals of the sample regression equation produces the following formulas for the estimated coefficients:

$$\hat{\beta}_1 = \frac{\sum_{i=1}^{n}(X_i - \bar{X})(Y_i - \bar{Y})}{\sum_{i=1}^{n}(X_i - \bar{X})^2}$$

REMEMBER

$$\hat{\beta}_0 = \bar{Y} - \hat{\beta}_1 \bar{X}$$

$\bar{X}$ is the mean (average) value of X; $\bar{Y}$ is the mean (average) value of Y.

As an example, suppose a medical researcher decides to test the relationship between the time spent working out in a gym and the corresponding weight loss. Eight volunteers spend a year working out in a local gym.

Table 15-1 shows the number of hours spent in the gym and corresponding weight loss (in pounds).

Table 15-1	Gym Time and Weight Loss
Y (Weight Loss in Pounds)	X (Hours Spent at the Gym)
15	100
11	75
15	80
14	90
8	60
9	50
2	25
5	40

Figure 15-1 shows a scatter plot for the data.

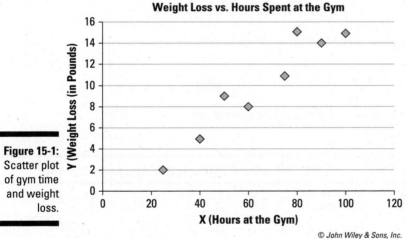

Figure 15-1: Scatter plot of gym time and weight loss.

© John Wiley & Sons, Inc.

The scatter plot shows that the relationship between gym time and weight loss is approximately *linear*. The scatter plot also shows that there's a positive (direct) relationship between weight loss and gym time; as gym time increases, weight loss also increases (and vice versa).

Because the relationship between X and Y appears to be linear, you can estimate the relationship between them with simple regression analysis.

If the relationship between X and Y is not linear, there are techniques available that may be able to transform the data into a linear form. For example, the dependent variable Y may be replaced with the natural log of Y, or $\ln(Y)$. You see this later in this chapter when we start to model financial returns.

In addition, there are specialized regression techniques that may be used with nonlinear data. These include *weighted least squares* (WLS) and *generalized least squares* (GLS). (These are beyond the scope of this book.)

The coefficients of the sample regression equation are computed as follows. The first step is to compute the sample means of X and Y. This is shown in the following equations:

$$\bar{X} = \frac{\sum_{i=1}^{n} X_i}{n} = \frac{X_1 + X_2 + X_3 + X_4 + X_5 + X_6 + X_7 + X_8}{n}$$

$$= \frac{100 + 75 + 80 + 90 + 60 + 50 + 25 + 40}{8} = 65$$

$$\bar{Y} = \frac{\sum_{i=1}^{n} Y_i}{n} = \frac{Y_1 + Y_2 + Y_3 + Y_4 + Y_5 + Y_6 + Y_7 + Y_8}{n}$$

$$= \frac{15 + 11 + 15 + 14 + 8 + 9 + 2 + 5}{8} = 9.875$$

Table 15-2 shows the remaining calculations.

Table 15-2	Computing the Regression Slope and Intercept				
Y (Weight Loss)	X (Hours at the Gym)	$(X_i - \bar{X})$	$(X_i - \bar{X})^2$	$(Y_i - \bar{Y})$	$(X_i - \bar{X})(Y_i - \bar{Y})$
15	100	35	1225	5.125	179.375
11	75	10	100	1.125	11.250
15	80	15	225	5.125	76.875
14	90	25	625	4.125	103.125
8	60	−5	25	−1.875	9.375
9	50	−15	225	−0.875	13.125
2	25	−40	1600	−7.875	315.000
5	40	−25	625	−4.875	121.875
Sum			**4650**		**830**

Compute the third column $(X_i - \bar{X})$ by subtracting the mean of X from each value of X. You compute the fourth column $(X_i - \bar{X})^2$ by squaring the value of $(X_i - \bar{X})$ in the third column. Compute the fifth column $(Y_i - \bar{Y})$ by subtracting the mean of Y from each value of Y. In the sixth column, $(X_i - \bar{X})(Y_i - \bar{Y})$ is computed by multiplying the third and fifth columns.

The sum in the fourth column shows that $\sum_{i=1}^{n} (X_i - \bar{X})^2 = 4650$.

The sum in the sixth column shows that $\sum_{i=1}^{n} (X_i - \bar{X})(Y_i - \bar{Y}) = 830$.

Based on these results, you compute the estimated coefficents as follows:

$$\hat{\beta}_1 = \frac{\sum_{i=1}^{n}(X_i - \bar{X})(Y_i - \bar{Y})}{\sum_{i=1}^{n}(X_i - \bar{X})^2} = \frac{830}{4650} = 0.1785$$

$$\hat{\beta}_0 = \bar{Y} - \hat{\beta}_1 \bar{X} = 9.875 - (0.1785)(65) = -1.7275$$

Based on these results, you write the estimated (sample) regression equation like this:

$$\hat{Y}_i = -1.7275 + 0.1785 X_i$$

The slope of this equation suggests that each additional hour of time spent at the gym leads to a weight loss of 0.1785 pounds. (Although the intercept of a regression equation doesn't always have an interpretation, in this case the intercept seems to indicate that someone who doesn't visit the gym will *gain* 1.7275 pounds!)

You can use the sample regression equation to predict the weight loss that would result from a specified amount of time spent working out. For example, for someone who works out for 50 hours, the sample regression equation predicts a weight loss of 7.1975 pounds:

$$\hat{Y}_i = -1.7275 + 0.1785(50) = 7.1975$$

Forecasting with a regression equation is valid only for values of X (hours at the gym) that fall within the range spanned by the sample data used to esti-mate the equation. In this example, the values of X range from 25 to 100 in the sample data. Therefore, this equation should only be used to forecast weight loss for working out times that range between 25 and 100 hours.

Testing the Estimated Regression Equation

Once you've estimated the regression equation, you need to test the results to be sure they're valid. That is, you need to see that the independent variable *(X)* really does explain the value of the dependent variable *(Y)*. You can check the overall fit of the model to the sample data with a measure known as the *coefficient of determination,* also known as R^2. The statistical validity of the regression coefficients can be tested with a hypothesis test based on the *Student's t-distribution;* this is known as the *t-test* (see Chapter 13 for more on the t-test).

The coefficient of determination (R^2)

The coefficient of determination, also known as R^2 ("*R*-squared"), is a statistical measure that shows the proportion of *variation* explained by the estimated regression line.

Variation refers to the sum of the squared *deviations* between the values of *Y* and the mean value of *Y*. Variation is given by the following formula:

$$\sum_{i=1}^{n}(Y_i - \bar{Y})^2$$

The smaller the variation of a regression model, the better the independent variable *(X)* explains the values of the dependent variable *(Y)*. In other words, the smaller is the variation, the better the regression model "fits" the sample data.

R^2 always takes on a value between 0 and 1, as shown in the following inequalities:

$$0 \le R^2 \le 1$$

The closer R^2 is to 1, the better the estimated regression equation fits or explains the relationship between *X* and *Y*.

Total sum of squares (TSS)

The expression

$$\sum_{i=1}^{n}(Y_i - \bar{Y})^2$$

is also known as the *total sum of squares* (TSS). This sum can be divided into two categories:

- Explained sum of squares (ESS)
- Residual sum of squares (RSS)

Explained sum of squares (ESS)

You compute the ESS with this formula:

$$ESS = \sum_{i=1}^{n}(\hat{Y}_i - \bar{Y})^2$$

This expression is also known as *explained variation*. It's the portion of total variation that measures how well the regression equation explains the relationship between X and Y.

Residual sum of squares (RSS)

You compute the RSS with this formula:

$$RSS = \sum_{i=1}^{n}(Y_i - \hat{Y}_i)^2$$

This expression is also known as *unexplained variation*. This is the portion of total variation that measures discrepancies (errors) between the actual values of Y and those estimated by the regression equation.

The sum of RSS and ESS equals TSS.

$$\sum_{i=1}^{n}(Y_i - \hat{Y}_i)^2 + \sum_{i=1}^{n}(\hat{Y}_i - \bar{Y})^2 = \sum_{i=1}^{n}(Y_i - \bar{Y})^2$$

That is, RSS + ESS = TSS.

Now, you can compute the coefficient of determination (R^2) in one of two equivalent ways:

$$R^2 = \frac{ESS}{TSS}$$

$$R^2 = 1 - \frac{RSS}{TSS}$$

For a simple regression equation containing a single dependent variable X, R^2 is the square of the correlation coefficient between X and Y. This means that the more closely X and Y track each other, the better the regression model will explain the relationship between X and Y.

Computing the coefficient of determination

Returning to the weight-loss example, you calculate the coefficient of determination.

Table 15-3 shows the calculations required for determining the value of R^2.

Table 15-3		Computing the Coefficient of Determination (R^2)			
Y_i	X_i	$\hat{Y}_i = -1.7275 + 0.1785X_i$	$RSS = \sum_{i=1}^{n}(Y_i - \hat{Y}_i)^2$	$ESS = \sum_{i=1}^{n}(\hat{Y}_i - \bar{Y})^2$	$TSS = \sum_{i=1}^{n}(Y_i - \bar{Y})^2$
15	100	16.123	1.260	39.031	26.266
11	75	11.660	0.436	3.186	1.266
15	80	12.553	5.990	7.169	26.266
14	90	14.338	0.114	19.914	17.016
8	60	8.983	0.965	0.797	3.516
9	50	7.198	3.249	7.169	0.766
2	25	2.735	0.540	50.980	62.016
5	40	5.413	0.170	19.914	23.766
Sum			**12.724**	**148.159**	**160.875**

Based on the results shown in Table 15-3:

$$R^2 = \frac{ESS}{TSS} = \frac{148.159}{160.875} = 0.9210$$

This shows that 92.10 percent of the variation in Y is *explained by* variation in X. The regression line does a good job of explaining the relationship between the two variables.

The t-test

Another important test of the results of regression analysis is to determine whether the slope coefficient (β_1) is different from zero. If the slope coefficient is close to zero, that indicates that X provides little or no explanatory power for the value of Y. In such a case, you should replace X with another independent variable in the regression equation.

To determine whether β_1 is different from zero, you conduct a *hypothesis test*. Chapter 5 introduces hypothesis testing, and examples of how to apply it appear throughout the chapters in Part III of this book. In this case, we use the *t-test* to determine whether or not the slope coefficient β_1 of the estimated regression equation is significantly different from zero. If $\beta_1 = 0$, that indicates that X does *not* explain the value of Y, and the regression results are then meaningless.

Null and alternative hypotheses

With the t-test, the null hypothesis is that the slope coefficient (β_1) equals zero

$$H_0: \beta_1 = 0$$

The alternative hypothesis is that the slope coefficient *does not* equal zero (null hypothesis rejected):

$$H_1: \beta_1 \neq 0$$

If the null hypothesis is accepted (if it cannot be rejected), then the independent variable X does *not* explain the value of the dependent variable Y.

Level of significance

When you test hypotheses about regression coefficients, the level of significance (α) is often set equal to 0.05 (5 percent). The level of significance is the probability of rejecting the null hypothesis when it is actually true.

Rejecting the null hypothesis when it is actually true is known as a *Type I error*. Failing to reject the null hypothesis when it is false is known as a *Type II error*. The smaller the level of significance (α), the lower the risk of committing a Type I error but the greater the risk of committing a Type II error. For many financial applications, a level of significance of 5 percent (0.05) is used.

Test statistic

For this type of hypothesis test, the test statistic is this:

$$t = \frac{\hat{\beta}_1}{s_{\hat{\beta}_1}}$$

This expression is known as a *t-statistic*, because it follows the t-distribution. The new term in this equation is as follows:

$s_{\hat{\beta}_1} =$ the *standard error* of $\hat{\beta}_1$

Think of the standard error of $\hat{\beta}_1$ as the standard deviation of $\hat{\beta}_1$. In other words, consider it to be the amount of uncertainty associated with using $\hat{\beta}_1$ to estimate β_1. You compute this in the following way:

The first step is to compute the *standard error of the estimate* (SEE). This is a measure of the dispersion of the sample values above and below the estimated regression line. Use this formula:

$$SEE = \sqrt{\frac{RSS}{n-2}} = \sqrt{\frac{\sum_{i=1}^{n}(Y_i - \hat{Y}_i)^2}{n-2}}$$

For the weight-loss example, based on the sample in Table 15-3 for the R^2 calculations, you get the following:

$$\sum_{i=1}^{n}(Y_i - \hat{Y}_i)^2 = 12.724$$

Therefore, with a sample size of 8,

$$SEE = \sqrt{\frac{\sum_{i=1}^{n}(Y_i - \hat{Y}_i)^2}{n-2}} = \sqrt{\frac{12.724}{6}} = 1.4563$$

The next step is to compute the *standard error* of $\hat{\beta}_1$. Here's the formula:

$$S_{\hat{\beta}_1} = \frac{SEE}{\sqrt{\sum_{i=1}^{n}X_i^2 - n\bar{X}^2}}$$

You can get the values in the denominator of this formula from Table 15-4.

Table 15-4	Computing the Standard Error of $\hat{\beta}_1$
Xi	X_i^2
100	10,000
75	5,625
80	6,400
90	8,100
60	3,600
50	2,500
25	625
40	1,600

The mean of X has previously been determined to be 65. The sum of the squared Xs equals:

$$\sum_{i=1}^{n} X_i^2 = 10,000 + 5,625 + 6,400 + 8,100 + 3,600 + 2,500 + 625 + 1,600 = 38,450$$

or

$$S_{\hat{\beta}_1} = \frac{SEE}{\sqrt{\sum_{i=1}^{n} X_i^2 - n\bar{X}^2}} = \frac{1.4563}{\sqrt{38450 - (8)(65^2)}} = 0.02136$$

With $\hat{\beta}_1 = 0.1785$, the t-statistic for $\hat{\beta}_1$ is computed as

$$t = \frac{\hat{\beta}_1}{s_{\hat{\beta}_1}} = \frac{0.1785}{0.02136} = 8.357$$

Critical values

The critical values are taken from the t-table with $n-2$ degrees of freedom. In the weight-loss example, the sample size was 8, so we use the t-distribution with 6 degrees of freedom.

In this case, say you choose the level of significance (α) to be 0.05. Since the sample size (n) is 8, the appropriate critical values are as follows:

$$\pm t_{\alpha/2}^{n-2} = \pm t_{0.025}^{6}$$

These are found in the t-table. Table 15-5 shows an excerpt.

Table 15-5		The t-distribution			
Degrees of Freedom	$t_{0.10}$	$t_{0.05}$	$t_{0.025}$	$t_{0.01}$	$t_{0.005}$
5	1.476	2.015	2.571	3.365	4.032
6	1.440	1.943	2.447	3.143	3.707
7	1.415	1.895	2.365	2.998	3.499
8	1.397	1.860	2.306	2.896	3.355
9	1.383	1.833	2.262	2.821	3.250
10	1.372	1.812	2.228	2.764	3.169

The value of the positive critical value $t_{0.025}^6$ (we're using a two-tailed test) is found at the intersection of the row with 6 degrees of freedom and the column labeled $t_{0.025}$. This equals 2.447. The value of the negative critical value $-t_{0.025}^6$ then is -2.447.

Decision rule

If the value of the test statistic is greater than 2.447, the null hypothesis $H_0: \beta_1 = 0$ is rejected in favor of the alternative hypothesis $H_0: \beta_1 > 0$. If the value of the test statistic is less than -2.447, then the null hypothesis $H_0: \beta_1 = 0$ is rejected in favor of the alternative hypothesis $H_0: \beta_1 < 0$. If the test statistic falls between -2.447 and 2.447, the null hypothesis $H_0: \beta_1 = 0$ is *not* rejected.

In this case, the test statistic is 8.357, which is greater than 2.447. Therefore, the null hypothesis $H_0: \beta_1 = 0$ is rejected in favor of $H_0: \beta_1 > 0$; this indicates that $\hat{\beta}_1$ is *statistically significant* (that is, different from zero). This is strong evidence that X (hours spent at the gym) *does* explain the value of Y (weight loss).

You can also test whether the estimated intercept ($\hat{\beta}_0$) is statistically significant, but this is usually not necessary because the slope coefficient is the most important value in the regression equation.

Using Statistical Software

Many spreadsheet programs and specialized statistical packages are available to generate the results you need for regression analysis. Using this kind of software avoids the need to do all the tedious calculations by hand!

Excel

For example, you can use a spreadsheet program such as Excel to generate the results for the weight-loss data. See Figure 15-2.

The printout shows the values of $\hat{\beta}_0$ and $\hat{\beta}_1$ under the Coefficients column. $\hat{\beta}_0$ is the intercept and has a value of -1.727150538, whereas $\hat{\beta}_1$ is the slope and has a value of 0.178494624.

Summary Output

Regression Statistics	
Multiple R	0.959637777
R Square	0.920904663
Adjusted R Square	0.907722107
Standard Error	1.456277353
Observations	8

Figure 15-2:
Excel print-
out of the
regression
results for
the weight-
loss data.

ANOVA

	df	SS	MS	F	Significance F
Regression	1	148.1505376	148.1505376	69.85782191	0.00015945
Residual	6	12.72446237	2.120743728		
Total	7	160.875			

	Coefficients	Standard Error	t Stat	P-value	Lower 95%	Upper 95%
Intercept	−1.727150538	1.480542224	−1.166566214	0.28765033	−5.349906852	1.895605777
X (Hours at the Gym)	0.178494624	0.021355887	8.35809918	0.00015945	0.12623865	0.230750597

© John Wiley & Sons, Inc.

The values of the coefficient of determination (R^2) and the standard error of the estimate (SEE) appear under the Regression Statistics column. R^2 ("R-square") has a value of 0.920904663, and the standard error of the estimate (shown as Standard Error in Figure 15-2) has a value of 1.456277353.

Using the p-value

The spreadsheet printout shown in Figure 15-2 provides one additional useful measure you can use to test hypotheses about the coefficients. These are *p-values* (that is, probability values). The values are under the P-value column. The p-value for the intercept ($\hat{\beta}_0$) is 0.28765033, and the p-value for the slope ($\hat{\beta}_1$) is 0.00015945.

The p-value represents the likelihood of obtaining the given t-statistic if the null hypothesis is true. An extremely low p-value indicates that the null hypothesis of a zero coefficient should be rejected. In fact, the p-value represents the lowest level of significance at which the null hypothesis can be rejected.

More formally, when testing the hypothesis H_0: $\beta_1 = 0$, if the p-value is less than the level of significance (α), the null hypothesis is rejected; otherwise, it is not rejected.

In this example, the p-value for $\hat{\beta}_1$ is 0.00015945; the level of significance is 0.05. Therefore, because the p-value is less than the level of significance, the null hypothesis H_0: $\beta_1 = 0$ is rejected, and your regression coefficient is statistically significant. This confirms the results found when testing the hypothesis with the t-statistic.

Using the t-statistic or the p-value to test the significance of a regression coefficient will always provide the same results.

Assumptions of Simple Linear Regression

The simple regression model shown in this chapter is based on several extremely important assumptions. If any of these assumptions is violated, the reliability of the regression results is questionable.

The most important assumptions are as follows:

- The expected value of each error term is zero: $E(\varepsilon_i) = 0$. This shows that although some error terms are positive and some are negative, on average they equal zero.

- The variances of the error terms are *finite* and *constant* for all values of x_i; this common variance is designated σ^2.

- The error terms are *independent* of each other (that is, they do not influence each other).

- Each error term ε_i is independent of the corresponding value of X_i.

- An assumption that isn't required for linear regression but is often used is that the error terms are *normally distributed*. Incorporating this assumption lets you compute confidence intervals for the regression coefficients. It also allows you to test hypotheses about the coefficients.

Violations of the assumptions

With simple regression analysis, two of the most important violations of these assumptions are autocorrelation and heteroscedasticity.

Autocorrelation

Autocorrelation occurs when the error terms are correlated with each other. This violates the assumption of independence.

Two independent variables have a correlation of *zero* (0) between them. The reverse is not necessarily true. It is possible to have uncorrelated variables that are not independent.

Autocorrelated error terms can cause the standard errors of the regression coefficients to be *understated*. This would increase the risk that coefficients will be found to be statistically significant when they are not.

Heteroscedasticity

Heteroscedasticity (that's easy for *you* to say!) occurs when the error terms don't have a constant variance. This problem can cause the standard errors of the regression coefficients to be understated, increasing the risk that coefficients will be found to be statistically significant when they are not.

Formal statistical tests are available to determine whether these problems are present. If they are, you must take corrective steps before continuing with regression analysis or choose a different modeling procedure.

Multiple Regression Analysis

With simple regression, you estimate a relationship between a dependent variable *(Y)* and an independent variable *(X)*. With multiple regression analysis, you estimate the relationship between a dependent variable *(Y)* and *two or more* independent variables $(X_1, X_2,...)$. This introduces several additional complexities, but also provides a great deal of additional flexibility, because you can include any number of variables in the regression equation.

Multiple regression uses the same basic approach as simple linear regression. We'll spare you the gory details in this section. The calculations involved in multiple regression can get unmanageable pretty quickly, so statistical software is used to build these multiple regression lines.

For example, the following multiple regression equation contains *k* independent variables:

$$Y_i = \beta_0 + \beta_1 X_{1i} + \beta_2 X_{2i} + ... + \beta_k X_{ki} + \varepsilon_i$$

For example, suppose a bank wants to determine whether the salaries of its non-managerial employees are closely related to education, experience, and hours worked per week. The bank chooses a random sample of eight employees. The annual salaries (measured in thousands of dollars per year), years of post-high school education, years of experience, and average hours worked each week are recorded.

Here are the following variables:

- ✔ Y represents an employee's annual salary, measured in thousands of dollars.

- ✔ X_1 represents an employee's years of education. A value of 0 represents someone who only has a high school diploma. Any value greater than 4 indicates graduate education.

- ✔ X_2 represents years of experience.

- ✔ X_3 represents average number of hours worked each week.

Table 15-5 shows the sample data.

Table 15-5	**Sample Data for Employees**		
Y (Annual Salary, $ Thousands)	X₁ (Years of Post-high School Education)	X₂ (Years of Experience)	X₃ (Average Hours Worked per Week)
82	4	4	40
48	0	2	40
60	4	1	50
85	6	1	50
72	0	4	50
62	0	3	40
90	4	3	50
101	6	4	60

The regression is run using Excel. Figure 15-3 shows the results.

Based on the printout in Figure 15-3, the estimated regression equation is as follows:

$$\hat{Y}_i = 18.558 + 4.945X_{1i} + 7.350X_{2i} + 0.450X_{3i}$$

(The coefficients are rounded to three digits past the decimal point.)

Summary Output

Regression Statistics	
Multiple R	0.948932566
R Square	0.900473014
Adjusted R Square	0.825827775
Standard Error	7.368308544
Observations	8

ANOVA

	df	SS	MS	F	Significance F
Regression	3	1964.832117	654.9440389	12.06336829	0.017944885
Residual	4	217.1678832	54.2919708		
Total	7	2182			

	Coefficients	Standard Error	t Stat	P-value	Lower 95%	Upper 95%
Intercept	18.55839416	21.72974066	0.854055023	0.441205151	−41.77303794	78.88982626
Education	4.945255474	1.40754299	3.513146095	0.024501853	1.037011984	8.853498955
Experience	7.350364964	2.269756121	3.238394158	0.031722164	1.048511691	13.65221824
Hours Worked	0.450364964	0.51528194	0.874016589	0.43144493	−0.980287056	1.881016984

Figure 15-3:
Printout
of salaries
regression
results.

© John Wiley & Sons, Inc.

This equation shows that at this bank, the following things are true:

- ✔ The starting salary for a new employee with no experience or post-high school education is $18,558. This is based on the intercept of the regression equation.

- ✔ Each additional year of college or graduate school adds $4,945 to an employee's salary; this is based on the coefficient of X_1.

- ✔ Each additional year of experience adds $7,350 to an employee's salary; this is based on the coefficient of X_2.

- ✔ Each additional hour worked per week adds $450 to an employee's salary; this is based on the coefficient of X_3.

Predicting the value of Y

This multiple regression equation can be used to *predict* the annual salary of an employee with a specific amount of experience and education. For example, suppose a randomly chosen employee is a college graduate, has 2 years of experience, and works an average of 40 hours per week. The predicted salary of this employee would be as follows:

$$\hat{Y}_i = 18.558 + 4.945X_{1i} + 7.350X_{2i} + 0.450X_{3i}$$

$$\hat{Y}_i = 18.558 + 4.945(4) + 7.350(2) + 0.450(40) = 71.038$$

This result shows that the predicted annual salary is $71.038 \times \$1,000$ ($\$71,038$).

You should do forecasting with a regression equation only with values contained within the sample data.

Testing the results of the multiple regression equation

Testing the results of a multiple regression equation is similar to testing the results of a simple regression equation. There are two main differences:

- ✔ You use the *adjusted* coefficient of determination instead of the coefficient of determination to indicate how well the regression equation fits the sample data.

- ✔ You use an additional hypothesis test to determine if *all* the slope coefficients are jointly equal to zero; this test is known as the *F-test*.

The adjusted coefficient of determination

The *adjusted coefficient of determination* is closely related to the coefficient of determination that's used with simple regression. The following equation shows the relationship between the adjusted coefficient of determination and the coefficient of determination:

$$\bar{R}^2 = 1 - (1 - R^2) \left[\frac{n-1}{n-(k+1)} \right]$$

In this equation:

$\bar{R}^2$ = The adjusted coefficient of determination (also known as "R bar squared")

R^2 = The coefficient of determination (also known as "R squared")

n = The sample size

k = The number of independent variables

As with the coefficient of determination, the range of possible values for the adjusted coefficient of determination is from 0 to 1:

$$0 \leq \bar{R}^2 \leq 1$$

Use the adjusted coefficient of determination with multiple regression analysis instead of the coefficient of determination. You do this because the coefficient of determination increases automatically as new independent variables are added to a regression equation, even if they *don't* contribute any new explanatory power to the equation.

The adjusted coefficient of determination, on the other hand, increases only when new independent variables are added that *do* increase the explanatory power of the regression equation. This makes it a much more useful measure of how well a multiple regression equation fits the sample data.

The adjusted coefficient of determination for the salaries data is found at the top of Figure 15-3. (It is listed as "Adjusted R Square.") This value equals approximately 0.8258. This means that the proportion of *variation* explained by the estimated regression line is 0.8258, or 82.58 percent.

The F-test

With a multiple regression equation, it's often useful to test the hypothesis that *all slope coefficients equal 0*. If this is true, then the regression equation does *not* explain the relationship between the dependent and the independent variables. In this case, a new set of independent variables may be used to try to explain the value of the dependent variable.

The null and alternative hypotheses

With three independent variables, this hypothesis test is set up as follows:

Here is the null hypothesis:

$$H_0: \beta_1 = \beta_2 = \beta_3 = 0$$

This hypothesis indicates that all three slope coefficients jointly equal 0. A coefficient of 0 indicates that an independent variable does *not* explain the value of the dependent variable. If this hypothesis cannot be rejected, then the regression equation cannot be used to explain the relationship between salaries, education, experience, and hours worked.

The alternative hypothesis is that at least one slope coefficient does *not* equal 0. In other words, at least one of the independent variables *does* belong in the regression equation.

The level of significance

In most business applications, you choose the level of significance to be 0.01, 0.05, or 0.10; 0.05 is a common choice. The Greek letter α (*alpha*) is often used to represent the level of significance.

The test statistic

You compute the test statistic (also known as the *F-statistic* because it follows the F-distribution, discussed in Chapter 13) as follows:

$$F = \frac{R^2 / k}{(1 - R^2) / [n - (k+1)]}$$

The key terms in this equation are as follows:

R^2 = Coefficient of determination

n = Sample size

k = Number of independent variables (that is, the number of Xs in the regression equation)

In this example, the F-statistic equals

$$F = \frac{R^2 / k}{(1 - R^2) / [n - (k+1)]} = \frac{0.9004 / 3}{(1 - 0.9004) / [8 - (3+1)]} = 12.05$$

The test statistic follows the *F-distribution* with k numerator degrees of freedom and $\left[n - (k+1) \right]$ denominator degrees of freedom. Figure 15-3 shows that the value of the F-statistic can also be found on the spreadsheet printout. The value of the F-statistic is found under the column headed "F"; this is 12.06 (the difference with the calculations above is due to rounding).

The critical values

In this case, there's a single critical value that is uniquely determined by the level of significance and two different types of degrees of freedom. As covered in Chapter 13, these are known as the *numerator* and *denominator* degrees of freedom. They are calculated as follows:

Numerator degrees of freedom = k = 3

Denominator degrees of freedom = $\left[n - (k+1) \right] = \left(8 - (3+1) \right) = 4$

You choose the appropriate critical value from an F-table. Unlike the tables used for most other probability distributions, one entire F-table is needed for a single value of the level of significance. Table 15-6 shows an excerpt from the F-table for a 5 percent level of significance ($\alpha = 0.05$).

Table 15-6	A Section of the F-table with $\alpha = 0.05$							
$v_2 \backslash v_1$	**2**	**3**	**4**	**5**	**6**	**7**	**8**	**9**
2	19.00	19.16	19.25	19.30	19.33	19.35	19.37	19.38
3	9.55	9.28	9.12	9.01	8.94	8.89	8.85	8.81
4	6.94	6.59	6.39	6.26	6.16	6.09	6.04	6.00
5	5.79	5.41	5.19	5.05	4.95	4.88	4.82	4.77
6	5.14	4.76	4.53	4.39	4.28	4.21	4.15	4.10
7	4.74	4.35	4.12	3.97	3.87	3.79	3.73	3.68
8	4.46	4.07	3.84	3.69	3.58	3.50	3.44	3.39
9	4.26	3.86	3.63	3.48	3.37	3.29	3.23	3.18

The top row represents the numerator degrees of freedom (v_1). The first column represents the denominator degrees of freedom (v_2).

v is the Greek letter often used to represent degrees of freedom.

In this example, you're looking for a right-tail area of 5 percent with $v_1 = 3$, and $v_2 = 4$. Remember from Chapter 13 that the F-distribution is one-tailed to the right, so all tests using the F-distribution are right-tailed tests. You find this critical value at the intersection of the column labeled 3 and the row labeled 4. This is shown as the following:

$$F_\alpha^{v1,v2} = F_{0.05}^{3,4} = 6.59$$

The decision rule

If the test statistic exceeds the critical value, the null hypothesis is rejected; otherwise, it is not. In this case:

✔ The test statistic is approximately 12.05.

✔ The critical value is 6.59.

Because the test statistic is greater than the critical value, the appropriate decision is to *reject* the null hypothesis $H_0: \beta_1 = \beta_2 = \beta_3 = 0$. This indicates that *at least* one of the slope coefficients is statistically significant at the 5 percent level. In other words, at least one of the variables (education, experience, or hours worked) *does* explain the annual salary of an employee at this company.

Testing hypotheses with the p-value

As an alternative to comparing the F-statistic with a critical value, you can test the hypothesis by comparing the p-value with the level of significance.

When you're using the p-value, the decision rule goes like this:

- ✔ If the p-value is less than the level of significance, reject the null hypothesis.
- ✔ Otherwise, do *not* reject the null hypothesis.

In the example, the level of significance is 0.05 (5 percent). Figure 15-3 shows the p-value (under the heading "Significance F") as approximately 0.0179. Because the p-value is well below the level of significance, the null hypothesis is rejected. Therefore, at least one of the slope coefficients is statistically significant at the 5 percent level.

The t-test

Once you have used the F-test to confirm that at least one slope coefficient isn't equal to 0, you test each slope coefficient separately using the t-test. The t-test lets you determine which of the slope coefficients is statistically significant, or if both are statistically significant.

Null and alternative hypotheses

With the t-test, the null hypothesis is that the slope coefficient equals 0. For example, you can test whether or not $\beta_1 = 0$ with the following null and alternative hypotheses:

$$H_0: \beta_1 = 0$$

$$H_1: \beta_1 \neq 0$$

The level of significance

When you test hypotheses about regression coefficients, you typically set the level of significance (α) to 0.05 (5 percent).

The test statistic

For this type of hypothesis test, the test statistic is the ratio of the estimated coefficient to the standard error of the coefficient. For example, the test statistic for determining if $\beta_1 = 0$ is this:

$$t = \frac{\hat{\beta}_1}{s_{\hat{\beta}_1}}$$

This expression is known as a *t-statistic,* because it follows the t-distribution.

You can find the values you need to construct the t-statistic from the regression printout under the Coefficients and Standard Error headings.

For X_1 (education), the coefficient is approximately 4.945, and the standard error is approximately 1.408. The t-statistic is computed as $4.945 / 1.408 = 3.512$. Similar calculations show that the t-statistics for X_2 (experience) and X_3 (hours worked) are 3.238 and 0.874, respectively.

The critical values

With a multiple regression equation, you take the critical values from the t-table with $n - (k + 1)$ degrees of freedom. n is the sample size, and k is the number of independent variables. In this case, $k = 3$ since there are three independent variables.

To test a hypothesis for each slope coefficient with a level of significance of 0.05, the appropriate critical values will be like this:

$$\pm t_{\alpha/2}^{n-(k+1)} = \pm t_{0.025}^{4}$$

Remember, we're using a two-tailed test here, which is why the subscript has the value .025 rather than .05. These values are in the t-table. Table 15-7 shows an excerpt.

Table 15-7		The t-distribution			
Degrees of Freedom	$t_{0.10}$	$t_{0.05}$	$t_{0.025}$	$t_{0.01}$	$t_{0.005}$
4	1.533	2.132	2.776	3.747	4.604
5	1.476	2.015	2.571	3.365	4.032
6	1.440	1.943	2.447	3.143	3.707
7	1.415	1.895	2.365	2.998	3.499
8	1.397	1.860	2.306	2.896	3.355
9	1.383	1.833	2.262	2.821	3.250
10	1.372	1.812	2.228	2.764	3.169

The value of the positive critical value $t_{0.025}^{4}$ is found at the intersection of the row labeled 4 degrees of freedom and the $t_{0.025}$ column. This equals 2.776. The value of the corresponding negative critical value is –2.776.

The decision rule

For testing the hypothesis $H_0: \beta_1 = 0$, you reach the appropriate decision as follows:

- ✔ If the value of the test statistic is greater than 2.776, the null hypothesis $H_0: \beta_1 = 0$ is rejected in favor of the alternative hypothesis $H_1: \beta_1 > 0$.

- ✔ If the value of the test statistic is less than –2.776, then the null hypothesis $H_0: \beta_1 = 0$ is rejected in favor of the alternative hypothesis $H_1: \beta_1 < 0$.

- ✔ If the test statistic falls between –2.776 and 2.776, the null hypothesis $H_0: \beta_1 = 0$ is *not* rejected.

You follow the same process when you test the hypothesis $H_0: \beta_2 = 0$ and the hypothesis $H_0: \beta_3 = 0$.

For β_1, the test statistic is 3.512, which is greater than 2.776. Therefore, the null hypothesis $H_0: \beta_1 = 0$ is rejected in favor of $H_1: \beta_1 > 0$. This indicates that β_1 is different from 0 (that is, it is *statistically significant*). Therefore, there's strong evidence that X_1 (education) helps explain the value of Y (annual salary).

For β_2, the test statistic is 3.238, which is greater than 2.776. Therefore, the null hypothesis $H_0: \beta_2 = 0$ is rejected in favor of $H_2: \beta_2 > 0$; this indicates that β_2 is different from 0 (that is, it is *statistically significant*). Therefore, there's strong evidence that X_2 (experience) explains the value of Y (annual salary).

For β_3, the test statistic is 0.874, which is between –2.776 and 2.776. Therefore, the null hypothesis $H_0: \beta_3 = 0$ is *not* rejected. Therefore, X_3 (hours worked) does *not* explain the value of Y (annual salary).

Testing hypotheses with the p-value

As an alternative of comparing the t-statistic with critical values, you can test the hypothesis by comparing the p-value with the level of significance. The decision rule is then as follows:

- ✔ If the p-value is less than the level of significance, reject the null hypothesis.

- ✔ Otherwise, do *not* reject the null hypothesis.

In this example, the level of significance is 0.05 (5 percent). The p-values for X_1, X_2, and X_3 are 0.0246, 0.0317, and 0.431, respectively. This shows that X_1 and X_2 are statistically significant, whereas X_3 is not.

Multicollinearity

One of the potential difficulties that comes with multiple regression analysis is *multicollinearity*. This occurs when two or more of the independent variables are highly correlated with each other.

Multicollinearity is unique to multiple regression. With a simple regression, there's only a single independent variable.

Multicollinearity causes the correlated variables to have large standard errors, so they appear to be statistically insignificant even if they are not.

Multicollinearity is indicated if the coefficient of determination R^2 exceeds 0.8, whereas the t-tests on individual coefficients show that they are not statistically significant.

One approach to removing multicollinearity is to eliminate one of the correlated variables from the regression. Doing so has the effect of lowering the p-values of the uncorrelated independent variables, which reduces the risk that they will be considered statistically insignificant when they are not.

We cover a lot of detail in this chapter. That's because regression models are widely used in all kinds of applications. The mechanics of building the models is handled pretty easily by statistical software. But it is important to know how to interpret what those software packages feed back to you. In particular, it's important to know how to evaluate how well your model is performing. You should continue to re-evaluate your regression models occasionally. Coefficients in these models can change over time.

Chapter 16

When You've Got the Time: Time Series Analysis

In This Chapter

▶ Understanding the properties of a time series

▶ Forecasting a time series with decomposition methods and smoothing techniques

▶ Understanding how a time series may be modeled and forecasted with regression analysis

A *time series* is a sequence of values observed over a period of time. For example, the daily closing price of IBM stock over the past 30 trading days is a time series. The monthly rainfall in Seattle over the past five years is also a time series, as is the daily price of gold over the past two years.

For many applications, a time series is the most useful organization of data. For example, the properties of a stock could be best analyzed with a time series of historical prices. As another example, you might produce a forecast of the demand for a company's products, based on the past buying patterns of the company's customers.

This chapter covers the key statistical features needed to analyze a time series, along with three forecasting techniques: decomposition methods, smoothing techniques, and time series regression.

Key Properties of a Time Series

The properties of a time series may be modeled in terms of the following components or factors. Most time series contain one or more of the following:

✔ Trend component

✔ Seasonal component

  ✔ Cyclical component
  ✔ Irregular component

Trend component

A *trend* is a long-run increase or decrease in a time series. As an example, gold prices over the past 40 years would show a very strong positive trend, as prices have risen consistently over this period. Other items that would show a long-run positive trend over recent decades would be the population of the United States, the Dow Jones Industrial Average, the size of the U.S. federal deficit, and so forth. In recent years, U.S. interest rates would show a strong negative trend since reaching a peak in 2000. From a modeling perspective, the trend is the most important component of a time series.

Seasonal component

Much economic data is affected by the time of the year. For example, sales of bathing suits, surfboards, swimming pools, gardening equipment, and similar items are much stronger during the warmer months. The demand for heating oil rises during the winter and falls during the warmer months. Sales of snow shovels, turkeys, and pumpkin pies are stronger during the colder months and weaker during the warmer months.

However, some items, such as cellphones, aren't heavily affected by the season. The seasonal component is often measured on a quarterly or monthly basis.

Cyclical component

The business cycle has an impact on virtually all economic activity. For example, sales of expensive items such as new homes, new cars, new furniture, and so forth typically decline when the economy falls into recession. As another example, sales at fast-food chains may rise during recessions, when consumers are more cost-conscious, and then fall during recoveries. On the other hand, the demand for basic consumer staples, such as shampoo and detergent, depends less on the business cycle.

The cyclical component is measured over a long time horizon, typically one year or longer.

Irregular component

Irregular effects are the impact of random events such as strikes, earthquakes, and sudden changes in the weather. By their nature, these effects are completely unpredictable.

Forecasting with Decomposition Methods

Decomposition methods are based on an analysis of the individual components of a time series. The strength of each component is estimated separately and then substituted into a model that explains the behavior of the time series. Two of the more important decomposition methods are

- Multiplicative decomposition
- Additive decomposition

Multiplicative decomposition

The *multiplicative decomposition* model is expressed as the product of the four components of a time series:

$$y_t = TR_t \, S_t \, C_t \, I_t$$

These variables are defined as follows:

y_t = Value of the time series at time t

TR_t = Trend at time t

S_t = Seasonal component at time t

C_t = Cyclical component at time t

I_t = Irregular component at time t

Each component has a subscript t to indicate a specific time period. The time period can be measured in weeks, months, quarters, years, and so forth.

For example, sales of air conditioners depend heavily on the season of the year; due to population growth, sales of air conditioners also show a positive trend over time. Suppose you use the following equation to estimate (and to explain) the trend in the demand for air conditioners:

$$TR_t = 1000 + 25t$$

Quarterly data is used, so t represents the time measured in quarters. This equation indicates that over time, sales of air conditioners tend to rise by 25 units per quarter. Using the trend equation, the forecast of air conditioner sales over the coming year looks like this:

$$TR_1 = 1000 + 25(1) = 1025$$
$$TR_2 = 1000 + 25(2) = 1050$$
$$TR_3 = 1000 + 25(3) = 1075$$
$$TR_4 = 1000 + 25(4) = 1100$$

Seasonal factors are handled by giving different weights to each season that are used to adjust the trend components. Assume that the seasonal factors for four seasons are as follows:

$$S_1 = 0.2$$
$$S_2 = 0.6$$
$$S_3 = 1.4$$
$$S_4 = 0.3$$

These values show that the seasonal demand for air conditioners is strongest in the third quarter and weakest in the fourth and first quarters. (If there is no seasonal effect, then each of these factors would be equal to 1.) Incorporating the seasonal factors into the model gives the following adjusted forecasts:

$$TR_1 S_1 = 1025(0.2) = 205$$
$$TR_2 S_2 = 1050(0.6) = 630$$
$$TR_3 S_3 = 1075(1.4) = 1505$$
$$TR_4 S_4 = 1100(0.3) = 330$$

Now, suppose you estimate the four cyclical (quarterly) factors to be:

$$C_1 = 0.90$$
$$C_2 = 1.10$$
$$C_3 = 1.15$$
$$C_4 = 1.05$$

Incorporating the cyclical factors gives the following adjusted forecast for the four quarters over the coming year:

$$TR_1 S_1 C_1 = 1025(0.2)(0.90) = 184.50$$
$$TR_2 S_2 C_2 = 1050(0.6)(1.10) = 693.00$$
$$TR_3 S_3 C_3 = 1075(1.4)(1.15) = 1730.75$$
$$TR_4 S_4 C_4 = 1100(0.3)(1.05) = 346.50$$

Additive decomposition

With *additive decomposition,* a time series is modeled as the *sum* of the trend, seasonal effect, cyclical effect, and irregular effects. This is shown in the following equation:

$$y_t = TR_t + S_t + C_t + I_t$$

The additive decomposition method is more appropriate when the seasonal factors tend to be steady from one year to the next. By contrast, multiplicative decomposition is more widely used since many economic time series have a seasonal factor that grows proportionately with the level of the time series. In other words, economic growth tends to be multiplicative rather than linear, because returns are compounded over time.

For example, sales of ice-cream increase more during the summer *as the population grows,* so the seasonal factor increases over time. In this case, you'd use multiplicative decomposition to forecast the demand for ice-cream.

Smoothing Techniques

Smoothing techniques remove fluctuations in a time series. The fluctuations may be due to seasonal and irregular components, so the result of removing them from the time series reflects only the trend and cyclical components. You can estimate the seasonal component (if any) separately.

Here are three widely used smoothing techniques:

- ✔ Moving averages
- ✔ Centered moving averages
- ✔ Exponential smoothing

You may recall that Chapter 5 discusses four data types: nominal, ordinal, interval, and ratio, and that not all statistical processes can be legitimately applied to all four data types. When applying smoothing techniques, these distinctions are important. All the smoothing techniques require interval data, and some of them, like exponential smoothing, require ratio data.

Moving averages

An *n*-period *moving average* is the average value of the *n* most recent observations taken from a time series. This is computed as follows:

$$\frac{y_t + y_{t-1} + \ldots + y_{t-n+1}}{n}$$

where

y_t is the value of the time series at time t

n is the size of the period

For example, Table 16-1 shows the prices for a stock over the past ten months.

Table 16-1	Sample Data for Stock Prices
Time	*Price*
October 2013	100
November 2013	101
December 2013	103
January 2014	99
February 2014	97
March 2014	102
April 2014	101
May 2014	98
June 2014	104
July 2014	106

A three-month moving average is constructed as follows. The average of the first three observations would be $(100 + 101 + 103)/3 = 101.33$. The average of the next three observations (the second, third, and fourth months) is $(101 + 103 + 99)/3 = 101$. You continue this process for the entire dataset. See Table 16-2 for the resulting three-month moving averages (MAs).

The first three-month moving average is listed next to November 2013, even though it represents the average of October 2013, November 2013, and December 2013. This shows that November 2013 is the "center" of the moving average.

Table 16-2	Three-Month Moving Average for Stock Prices	
Time	*Price*	*3-Month MA*
October 2013	100	***
November 2013	101	101.33
December 2013	103	101.00
January 2014	99	99.67
February 2014	97	99.33
March 2014	102	100.00
April 2014	101	100.33
May 2014	98	101.00
June 2014	104	102.67
July 2014	106	***

Similarly, the three-month moving average constructed from the November 2013, December 2013, and January 2014 prices is shown next to December 2013. Plotting these moving averages and the original prices shows that moving averages "smooth out" the data, as shown in Figure 16-1.

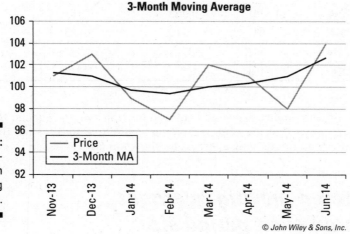

Figure 16-1: Three-month moving average.

© John Wiley & Sons, Inc.

The size of the period used to construct a moving average is often determined by the characteristics of the data. For example, 12-month moving averages are often used with monthly data.

Centered moving averages with an odd period size

A *centered moving average* is an average of moving averages. For example, using the stock prices from the previous example (Table 16-1), the first three-month moving average is 101.33, and the second three-month moving average is 101. The centered moving average is $(101.33 + 101)/2 = 101.67$.

See Table 16-3 for the resulting centered moving averages.

Table 16-3	Three-Month Centered Moving Average for Stock Prices		
Time	*Price*	*3-Month MA*	*3-Month Centered MA*
October 2013	100	***	***
November 2013	101	101.33	101.17
December 2013	103	101.00	100.33
January 2014	99	99.67	99.50
February 2014	97	99.33	99.67
March 2014	102	100.00	100.17
April 2014	101	100.33	100.67
May 2014	98	101.00	101.83
June 2014	104	102.67	***
July 2014	106	***	***

Plotting the moving average and centered moving average against the original prices shows that the centered moving average smoothes out the data even more than the moving average, as shown in Figure 16-2.

Centered moving averages with an even period size

If the size of the period is an even number, the computation of a centered moving average is slightly different than it is for an odd number. For example, Table 16-4 shows the results of computing a four-month centered moving average for the same data.

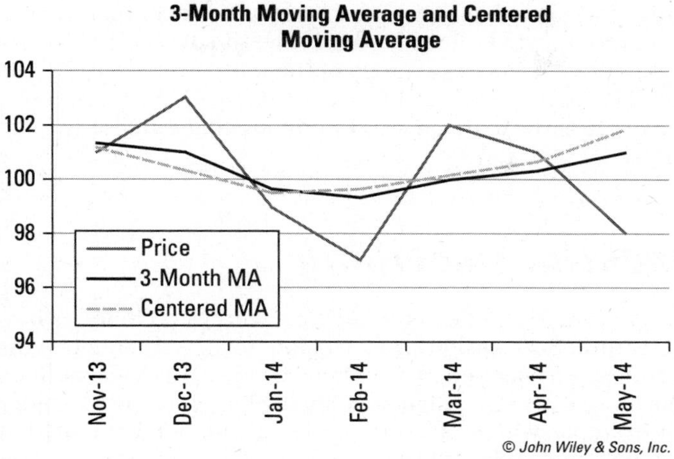

3-Month Moving Average and Centered Moving Average

Figure 16-2:
Three-month moving average and centered moving average.

© John Wiley & Sons, Inc.

Table 16-4		Four-Month Centered Moving Average for Stock Prices		
Time	**Price**	**4-Month Moving Total**	**Average 4-Month Moving Total**	**4-Month Centered MA**
October 2013	100	***	***	***
November 2013	101	***	***	***
December 2013	103	403.00	***	***
January 2014	99	400.00	401.50	100.38
February 2014	97	401.00	400.50	100.13
March 2014	102	399.00	400.00	100.00
April 2014	101	398.00	398.50	99.63
May 2014	98	405.00	401.50	100.38
June 2014	104	409.00	407.00	101.75
July 2014	106	***	***	***

The four-month moving total is the sum of the prices for the previous two months, the current month, and the following month. For example, the four-month moving total for December 2013 is the sum of the prices for October 2013, November 2013, December 2013, and January 2014.

The average four-month moving total is the average of the previous and current four-month moving totals. For example, the average four-month moving total for January 2014 is the average of the four-month moving totals for December 2013 and January 2014. For each month, the four-month centered moving average is the average four-month moving total for the same month divided by 4.

Exponential smoothing

The moving average and centered moving average techniques have one common feature: Both assign equal weights to all elements of a time series. If a time series consists of data that becomes less relevant as time elapses, it may make more sense to assign steadily declining weights to older observations. For example, with stock prices, the information contained in recent prices is much more relevant than the information contained in older prices, so it would make sense to develop a time series model based on declining weights. You can do that with *exponential smoothing*.

With exponential smoothing, the members of the time series are *weighted* in such a way as to ensure that newer observations assume more importance than older observations. You implement the scheme using *smoothing constants*.

The exponential smoothing approach uses the following formula:

$$E_t = \alpha y_t + (1 - \alpha)E_{t-1}$$

where

E_t = Exponentially smoothed value at time t

E_{t-1} = Exponentially smoothed value at time $t - 1$ (that is, one period in the past)

α = The smoothing constant — α assumes a value between 0 and 1

y_t = The value of the time series at time t

For example, Table 16-5 shows daily gold prices between 7/23/14 and 8/5/2014.

Table 16-5	Daily Gold Prices
Date	*Price ($/Ounce)*
7/23/2014	1,304.51
7/24/2014	1,293.54

Date	Price ($/Ounce)
7/25/2014	1,306.98
7/28/2014	1,303.95
7/29/2014	1,299.09
7/30/2014	1,296.12
7/31/2014	1,282.51
8/1/2014	1,293.66
8/4/2014	1,288.13
8/5/2014	1,288.47

Table 16-6 shows the exponentially smoothed values of the daily gold prices; in the third column, α is set equal to 0.3, and in the fourth column, α is set equal to 0.7.

Table 16-6	Exponentially Smoothed Daily Gold Prices		
Date	Price ($/Ounce)	Exponentially Smoothed Price $(\alpha = 0.3)$	Exponentially Smoothed Price $(\alpha = 0.7)$
7/23/2014	1,304.51	1,304.51	1,304.51
7/24/2014	1,293.54	1,301.22	1,302.21
7/25/2014	1,306.98	1,302.95	1,302.73
7/28/2014	1,303.95	1,303.25	1,303.09
7/29/2014	1,299.09	1,302.00	1,302.33
7/30/2014	1,296.12	1,300.24	1,300.86
7/31/2014	1,282.51	1,294.92	1,296.70
8/1/2014	1,293.66	1,294.54	1,295.19
8/4/2014	1,288.13	1,292.62	1,293.39
8/5/2014	1,288.47	1,291.37	1,291.98

You perform the calculations for $\alpha = 0.3$ as follows:

For 7/23/14, the exponentially smoothed price equals the actual price of $1,304.51. For 7/24/14, the exponentially smoothed price equals

$(0.30 \times 7/24/14 \text{ price}) + (0.70 \times 7/23/14 \text{ exponentially smoothed}$

$\text{price}) = (0.30 \times 1,293.54) + (0.70 \times 1,304.51) = 1,301.22.$

For 7/25/14, the exponentially smoothed price equals $(0.30 \times$ the 7/25/14 price$) + (0.70 \times 7/24/14$ exponentially smoothed price$) = 0.30 \times 1,306.98) + (0.70 \times 1,301.22) = 1,302.95$.

The same procedure is used for $\alpha = 0.7$. In general, the exponentially smoothed price equals

$(\alpha \times$ current price$) + ((1-\alpha) \times$ previous day's exponentially smoothed price$)$

Figure 16-3 shows the relationship between the actual gold prices and the exponentially smoothed gold prices.

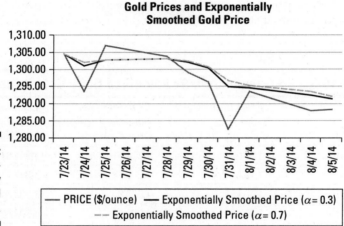

© John Wiley & Sons, Inc.

Figure 16-3: Exponentially smoothed gold prices.

The graph shows that the exponentially weighted values do not fluctuate as much as the original values, and that the values fluctuate less when $\alpha = 0.7$ than when $\alpha = 0.3$.

With an exponential smoothing model, you can make a *forecast* for the next period. The forecast for time $t+1$ as of time t equals:

$$F_{t+1} = \alpha y_t + (1-\alpha)F_t$$

where

F_t is the forecast for time t

F_{t+1} is the forecast for time $t+1$

For the gold price example, the price on 8/5/14 is \$1,288.47, while the exponentially smoothed price (forecast) for 8/6 is \$1,291.37. Here's the calculation:

$$F_{t+1} = (0.30)(1,288.47) + (0.70)(1,291.37) = 1,290.50$$

Seasonal Components

Many time series contain strong seasonal components. In these cases, you can estimate the seasonal component of a time series with a technique known as the *ratio-to-moving average method*.

For example, Table 16-7 shows the quarterly sales (in millions of dollars) of ice-cream for a national grocery store chain from 2011–2014.

Table 16-7	Quarterly Ice-Cream Sales (\$ Millions)	
Year	*Quarter*	*Sales*
2011	1	14
2011	2	32
2011	3	46
2011	4	18
2012	1	16
2012	2	38
2012	3	48
2012	4	17
2013	1	16
2013	2	35
2013	3	49
2013	4	22
2014	1	15
2014	2	37
2014	3	52
2014	4	23

The seasonal components are derived from a four-month centered moving average, as shown in Table 16-8.

Table 16-8   **Calculation of Seasonal Components for Ice-Cream Sales**

Year	Quarter	Sales	4-Quarter Moving Total	Average 4-Quarter Moving Total	4-Quarter Centered MA	Sales as a % of 4-Quarter Centered MA
2011	1	14	***	***	***	***
2011	2	32	***	***	***	***
2011	3	46	110	***	***	***
2011	4	18	112	111.00	27.75	64.8604
2012	1	16	118	115.00	28.75	55.6504
2012	2	38	120	119.00	29.75	127.7304
2012	3	48	119	119.50	29.88	160.6704
2012	4	17	119	119.00	29.75	57.1404
2013	1	16	116	117.50	29.38	54.4704
2013	2	35	117	116.50	29.13	120.1704
2013	3	49	122	119.50	29.88	164.0204
2013	4	22	121	121.50	30.38	72.4304
2014	1	15	123	122.00	30.50	49.1804
2014	2	37	126	124.50	31.13	118.8804
2014	3	52	127	126.50	31.63	164.4304
2014	4	23	***	***	***	***

The average sales percentage for each quarter is as follows:

     Q1 : 53.10%   Q2 : 122.2604%

     Q3 : 163.04%   Q4 : 64.8104%

You make one final adjustment. The sum of these percentages is 403.2104%. You want the average adjustment to be 1 (or 100%). So, you apply an adjustment factor of 400% / 403.21% (which results in 99.20%) to each of the average sales percentages to get the appropriate seasonal components:

     Q1 : 52.6804%   Q2 : 121.2904%

     Q3 : 161.7404%   Q4 : 64.3004%

These can be substituted into the decomposition model like so:

$$S_1 = 0.5268 \qquad S_2 = 1.2129$$
$$S_3 = 1.6174 \qquad S_4 = 0.6430$$

Once you've computed the seasonal components, you can remove the influence of seasonal variation from a time series. This process is known as *deseasonalizing* the time series. A time series that has been deseasonalized contains *seasonally adjusted* data (a term you've no doubt frequently heard on TV), making it possible to analyze just the trend and cyclical components of a time series.

For the multiplicative decomposition method

$$y_t = TR_t \, S_t \, C_t \, I_t$$

the deseasonalized data is obtained by dividing out the seasonal component, like this:

$$y_t \, / \, S_t = TR_t \, C_t \, I_t$$

For the additive decomposition method

$$y_t = TR_t + S_t + C_t + I_t$$

the deseasonalized data is obtained by subtracting the seasonal component:

$$y_t - S_t = TR_t + C_t + I_t$$

Modeling a Time Series with Regression Analysis

An alternative approach to modeling the components of a time series is to use *regression analysis* (Chapter 15 covers regression analysis).

The basic form of a time series regression model goes like this:

$$y_t = TR_t + e_t$$

The key terms in this equation are as follows:

TR_t = The trend of the time series at time t

ε_t = An error term at time t

You can model seasonal components and cyclical components by including additional terms in the regression equation.

Identifying the trend

The first step in estimating a time series regression model is to identify the type of trend (if any) that's present in the data. The basic types of trends that may appear in a time series include the following:

- ✔ No trend
- ✔ Linear trend
- ✔ Quadratic trend
- ✔ Exponential trend

No trend

When there's no trend, the values of the time series may rise or fall over time, but on average they tend to return to the same level. The trend is written like this:

$$TR_t = \beta_0$$

Figure 16-4 shows a time series with no trend.

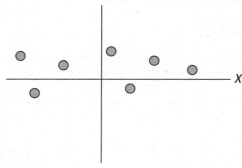

Figure 16-4:
A time series with no trend.

© John Wiley & Sons, Inc.

Linear trend

With a linear trend, the values of a time series tend to rise or fall over time at a constant rate (β_1). If the values tend to rise, then the time series has a positive trend; if the values tend to fall, then the time series has a negative trend. For a linear trend:

$$TR_t = \beta_0 + \beta_1 t$$

Figure 16-5 shows a time series with a positive linear trend.

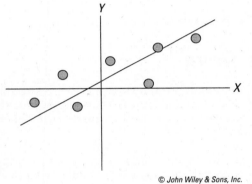

Figure 16-5:
A time series with a positive linear trend.

© John Wiley & Sons, Inc.

Quadratic trend

With a quadratic trend, the values of a time series tend to rise or fall at a rate that changes over time. You model a quadratic trend using the following equation:

$$TR_t = \beta_0 + \beta_1 t + \beta_2 t^2$$

Figure 16-6 shows a time series with a quadratic trend. In this case, the value of y_t increases at an increasing rate over time.

Exponential trend

This trend is used when the growth rate of a time series is proportional to the current level of the time series. In other words, growth is faster at higher values of the time series than at lower values. The general expression for a higher order polynomial trend goes like this:

$$TR_t = e^{(\beta_0 + \beta_1 t)}$$

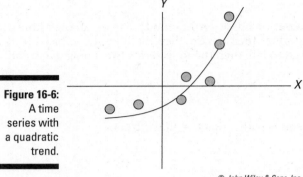

Figure 16-6:
A time series with a quadratic trend.

© John Wiley & Sons, Inc.

Estimating the trend

The first step in estimating the trend of a time series is to create a time series plot to analyze the characteristics of the data. The time series plot shows whether there's a trend in the data; if there is one, you can use the plot to identify the order of the trend.

For example, suppose a portfolio manager has reason to believe that there may be a linear trend in a time series of daily ExxonMobil stock prices from August 2004 to August 2014. These are shown in Figure 16-7.

Daily Prices of ExxonMobil Stock
August 2004 – August 2014

Figure 16-7:
Daily prices for ExxonMobil stock.

© John Wiley & Sons, Inc.

In this case, there does appear to be a trend in the data; although the price does fluctuate, over time it is steadily rising.

To formally test whether there's a linear trend in the data, you can run a time series regression with a time trend as the independent variable; this is written as follows:

$$y_t = \beta_0 + \beta_1 t + \varepsilon_t$$

In this example

✔ The dependent variable (y_t) is the price of ExxonMobil stock.

✔ The independent variable (t) is time (measured in months).

Figure 16-8 shows the results of running a regression of the price of ExxonMobil stock against time.

Summary Output

Regression Statistics	
Multiple R	0.85340554
R Square	0.72830101
Adjusted R Square	0.72819306
Standard Error	8.14413406
Observations	2519

ANOVA

	df	SS	MS	F	Significance F
Regression	1	447502.9788	447502.979	6746.92842	0
Residual	2517	166944.8565	66.3269195		
Total	2518	614447.8353			

	Coefficients	Standard Error	t Stat	P-value	Lower 95%	Upper 95%
Intercept	45.1244604	0.324631116	139.002265	0	44.48788897	45.7610318
Time	0.01832932	0.000223148	82.1396885	0	0.017891748	0.01876689

© John Wiley & Sons, Inc.

Figure 16-8: Regression results for ExxonMobil stock: linear trend.

In this case, the independent variable is time. The first observation in the sample (8/2/2004) is assigned a time value of 1, the second observation in the sample (8/3/2004) is assigned a time value of 2, and so forth. The final observation on 8/1/2014 is assigned a time value of 2,519.

In the regression, the coefficient of time is approximately 0.0183, indicating that on average the price of ExxonMobil stock rises by $0.0183 each trading day. That represents a positive trend. Here's the equation of the trend:

$$TR_t = \beta_0 + \beta_1 t$$

$$TR_t = 45.1244 + 0.0183t$$

The coefficient of time is statistically significant at the 5 percent level, because the p-value of the time coefficient is well below 0.05. (See Chapter 15 for more information.)

Forecasting with time series regression

You can use the time series regression equation to forecast future stock prices for ExxonMobil. For example, suppose a portfolio manager wants to forecast the price of ExxonMobil stock for August 2, 2014. Because the price on August 1, 2014, has a time value of 2,519, the time value for August 2, 2014, is 2,520. You substitute this into the trend equation to come up with a forecast for the price of ExxonMobil stock on August 2, 2014:

$$45.1244 + 0.0183t = 45.1244 + 0.0183(2,520) = 91.24$$

Prediction interval

A value predicted by a regression equation is subject to uncertainty; to estimate the range of possible values for a prediction, you can construct a *prediction interval*. The prediction interval is computed like so:

$$\hat{y}_t \pm t_{\alpha/2}^{n-1} s \sqrt{1 + \left(\frac{1}{n}\right) + \frac{(n+1-\bar{t})^2}{\sum_{t=1}^{n}(t-\bar{t})^2}}$$

In the ExxonMobile example, a 95 percent prediction interval is estimated. This indicates that $(1-\alpha) = 0.95$ so that $\alpha = 0.05$ and $\alpha/2 = 0.025$. Based on Figure 16-8, the standard error of the regression (s) equals 8.1441, and the sample size (n) equals 2,519.

The predicted value of ExxonMobil stock for 8/2/2014 is 91.24. The average value of t during the sample period is 1,260. The sum of the squared differences between t and the average value of t is 1,331,996,820.

Because the sample size is extremely large (you had more than 2,500 days of data), you can take the critical value from the standard normal distribution (z); for a 95 percent prediction interval, this value is 1.96. Substituting these results into the equation gives the following:

$$91.24 \pm (1.96)(8.1441)\sqrt{1 + \left(\frac{1}{2519}\right) + \frac{(2519 + 1 - 1260)^2}{1,331,996,820}}$$

$$= (75.28,\ 107.20)$$

The predicted value for ExxonMobil stock is $91.24. The prediction interval shows with 95 percent certainty that the actual value of ExxonMobil stock is expected to be between $75.28 and $107.20.

Estimating a quadratic trend

Now let's say the portfolio manager wants to determine whether there's a quadratic trend in the time series of ExxonMobil stock prices. To do so, he or she would use the following model:

$$TR_t = \beta_0 + \beta_1 t + \beta_2 t^2$$

Figure 16-9 shows the results of running this regression.

Summary Output

Regression Statistics	
Multiple R	0.855847606
R Square	0.732475125
Adjusted R Square	0.732262466
Standard Error	8.082938523
Observations	2519

ANOVA

	df	SS	MS	F	Significance F
Regression	2	450067.7551	225033.878	3444.36647	0
Residual	2516	164380.0802	65.3338952		
Total	2518	614447.8353			

	Coefficients	Standard Error	t Stat	P-value	Lower 95%	Upper 95%
Intercept	47.38344297	0.483527694	97.9953033	0	46.43529	48.331596
Time	0.012952924	0.000886216	14.6159962	1.6932E-46	0.01121514	0.01469071
Time Sq.	2.13349E-06	3.40514E-07	6.26549615	4.3578E-10	1.4658E-06	2.8012E-06

© *John Wiley & Sons, Inc.*

Figure 16-9: Regression results for ExxonMobil stock: quadratic trend.

Figure 16-9 shows that the coefficients of t and t^2 are both statistically significant, indicating that there is a *quadratic* trend in the data. The estimated model is this:

$$TR_t = 47.383 + 0.01295t + 0.000002133t^2$$

Here, then, is the predicted value for ExxonMobil stock on August 2, 2014:

$$47.383 + 0.01295(2,520) + 0.000002133(2,520)^2 = 93.57$$

Seasonal variation

As mentioned earlier, seasonal variation refers to recurring changes in a time series that are due to the season of the year. For example, the demand for petroleum products (gasoline and diesel fuel) tends to be greater during the summer than during the winter.

You can use a scatter plot to determine whether a time series exhibits seasonal variation. In this case, the time series regression model can be extended to include a variable (S_t) that represents the seasonal component:

$$y_t = TR_t + S_t + \varepsilon_t$$

Modeling seasonal variation with dummy variables

One method you can use to determine the effect of seasonality (S_t) is called a dummy variable. A *dummy variable* is a binary variable with a value of 0 or 1. A dummy variable is used to indicate whether a condition is true or false; a value of 1 indicates that the condition is true, while a value of 0 indicates that the condition is false.

You use a dummy variable to measure the effect of non-quantitative variables, such as gender, geographical region, and so forth, on the dependent variable. For modeling seasonal variation, you use a dummy variable to indicate whether an observation in a time series belongs to a given season.

For example, suppose an analyst wants to see whether the demand for heating oil depends on the quarter of the year. He or she has reason to believe that demand is highest in the fourth and first quarters due to the cold weather. The following seasonal dummy variables are defined:

$D_1 = 1$ if time period t is in the first quarter, 0 otherwise

$D_2 = 1$ if time period t is in the second quarter, 0 otherwise

$D_3 = 1$ if time period t is in the third quarter, 0 otherwise

In this case, there are only three dummy variables, not four. That's because including one for each season would lead to the problem of *multicollinearity*, which affects the reliability of the regression results. (Chapter 15 talks more about multicollinearity.) The fourth quarter is represented by process of elimination. If all three dummy variables are 0, then you're in the fourth quarter. As a result, the seasonal effect of the fourth quarter is captured by the trend.

With three seasonal dummy variables and a linear trend, the time series regression equation is as follows:

$$y_t = \beta_0 + \beta_1 t + \beta_2 D_1 + \beta_3 D_2 + \beta_4 D_3 + \varepsilon_t$$

Table 16-9 shows the relationship between the values of D_1, D_2, and D_3 and the quarter of the year.

Table 16-9	Dummy Variables		
Quarter	D_1	D_2	D_3
1	1	0	0
2	0	1	0
3	0	0	1
4	0	0	0

For example, imagine a department store that sells surfboards. Its sales are heavily dependent on the quarter of the year. In particular, sales are strongest during the second and third quarters (April–September), and are extremely weak during the first and fourth quarters (January–March and October–December).

To analyze the relationship between sales and the season, you run a regression with ten years of quarterly data. Sales is the dependent variable; the independent variables are a time trend and a series of three quarterly dummy variables.

These are defined as follows:

$D_1 = 1$ if first quarter, 0 otherwise

$D_2 = 1$ if second quarter, 0 otherwise

$D_3 = 1$ if third quarter, 0 otherwise

To avoid the problem of multicollinearity, no dummy variable is used for the fourth quarter. Instead, the intercept and the trend of the regression equation are used to capture the effect of the fourth quarter on sales. Figure 16-10 shows quarterly sales (in thousands of dollars) for 2005–2014. There's also a trend line.

The graph in Figure 16-10 shows that the trend line by itself does a very poor job of explaining sales because sales are highly seasonal. But when you run a regression with a trend and the seasonal dummies, you see the results shown in Figure 16-11.

The results show that each of the independent variables has a statistically significant coefficient and therefore belongs in the regression equation, because in each case the p-value is below 0.05.

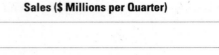

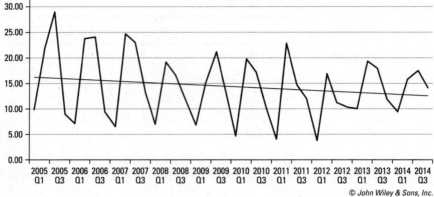

Figure 16-10:
Quarterly
sales.

© John Wiley & Sons, Inc.

Summary Output

Regression Statistics	
Multiple R	0.883631708
R Square	0.780804995
Adjusted R Square	0.755754138
Standard Error	3.157746335
Observations	40

Figure 16-11:
Time trend
regres-
sion with
seasonal
dummy
variables
for quarterly
sales.

ANOVA

	df	SS	MS	F	Significance F
Regression	4	1243.181258	310.795314	31.1687929	4.27358E-11
Residual	35	348.997667	9.97136191		
Total	39	1592.178925			

	Coefficients	Standard Error	t Stat	P-value	Lower 95%	Upper 95%
Intercept	13.90290102	1.382453507	10.0566861	7.3109E-12	11.09637121	16.7094308
t	-0.10708641	0.043457054	-2.46418938	0.01878846	-0.195308919	-0.0188639
D1	-4.855956296	1.418192167	-3.42404676	0.0015894	-7.73503944	-1.97687315
D2	8.246375747	1.414859161	5.8284075	1.2996E-06	5.374058965	11.1186925
D3	7.657160168	1.412855583	5.41963401	4.4985E-06	4.788910866	10.5254095

© John Wiley & Sons, Inc.

The (approximate) coefficients of the variables are as follows:

Intercept:	13.9029
Trend:	-0.1071
Dummy 1:	-4.8560
Dummy 2:	8.2464
Dummy 3:	7.6572

The trend indicates that sales are decreasing by $107.10 (0.1071 × $1,000) per quarter. The coefficients of the remaining dummy variables show the value of sales compared with a trend line at the level of average fourth quarter sales. This is a line with an intercept of 13.9029 and a slope of –0.1071.

The coefficient of D_1 shows that sales during the first quarter are below the trend by $4,856.00 (4.8560 × $1,000). The coefficient of D_2 shows that sales during the second quarter are above the trend by $8,246.40 (8.2464 × $1,000). The coefficient of D_3 shows that sales during the third quarter are above the trend by $7,657.20 (7.6572 × $1,000).

Cyclical variation

Cyclical variation refers to recurring changes in a time series that are due to the business cycle. For example, the demand for luxury goods like expensive cars tends to fall when the economy is in recession and rise when the economy recovers. In this case, the time series regression model can be extended to include a variable that represents the cyclical component (C_t):

$$y_t = TR_t + C_t + S_t + \varepsilon_t$$

This variable may also take the form of a series of dummy variables.

Comparing Different Models: MAD and MSE

Because you might use many different models to describe the behavior of a time series, it is important to be able to determine which model best "fits" or explains the observed data. The following are two techniques that may be used for this purpose:

- ✔ Mean absolute deviation (MAD)
- ✔ Mean square error (MSE)

Mean absolute deviation (MAD)

Mean absolute deviation (MAD) is computed as follows:

$$MAD = \frac{\sum_{t=1}^{n} |y_t - \hat{y}_t|}{n}$$

MAD is the average absolute value of the differences between the actual values of y_t and the predicted values.

Mean square error (MSE)

Mean square error (MSE) is computed as follows:

$$MSE = \frac{\sum_{t=1}^{n}(y_t - \hat{y}_t)^2}{n}$$

MSE is the average squared value of the differences between the actual values of y_t and the predicted values.

These measures make it possible to choose between different time series models based on their predictive ability. The lower the value of MAD or MSE, the more accurate a time series model is. These measures are clearly related because they use the same inputs. MSE can be thought of as a measure of how well your regression model fits the data, whereas MAD gives you a better sense of how close you are to the actual values you're trying to predict.

Part IV
Big Data Applications

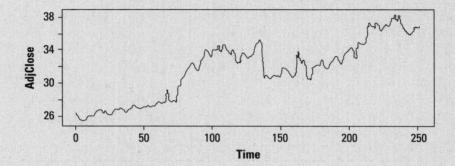

In this part . . .

- ✔ Seeing the future with forecasting
- ✔ Crunching the numbers on your desktop or laptop
- ✔ Finding sources of data online

Chapter 17

Using Your Crystal Ball: Forecasting with Big Data

● ●

In This Chapter

▶ Understanding the properties of time series

▶ Transforming data to fit modeling assumptions

▶ Forecasting a time series using ARIMA modeling

▶ Seeing how simulation is used for forecasting purposes

● ●

*Y*ou can use a few different techniques to forecast the future values of a time series:

✔ Time series regression

✔ ARIMA modeling

✔ Simulation

Chapter 16 covers time series regression. This chapter covers ARIMA modeling and simulation techniques. ARIMA models use the past values of a time series to develop a forecasting model, whereas simulation techniques are based on a statistical model of the variable that's being forecast.

ARIMA Modeling

ARIMA *(autoregressive integrated moving average)* modeling uses the past behavior of a time series to determine its key statistical properties and takes this information to develop a forecasting model.

ARIMA modeling is only valid for a time series that's both *stationary* and *nonseasonal*. A time series is stationary if the basic statistical properties of the time series don't change over time. In particular, this means that neither the values of the time series nor their variances grow (or shrink) over time. A time series is nonseasonal if its values are not affected by the time of the year. For example, if a time series consists of the sales of skis by a department store over a period of ten years, the data would show large increases in sales during winter months followed by significant declines during the rest of the year. As a result, you couldn't apply ARIMA modeling to this time series — because it's seasonal.

Testing for stationarity

There are two main ways you can evaluate a time series as nonstationary:

- ✔ Seeing whether the time series contains a *trend*
- ✔ Seeing whether the variance of the time series changes over time

A *trend* shows a long-term tendency for a variable to rise or fall. For example, the population of the United States shows a long-term positive trend.

You make a plot of a time series to see whether there's any indication of *nonstationarity*. If the plot shows a trend (positive or negative), this is a sign that the time series may be *nonstationary*. If the plot shows increasingly large (or small) fluctuations over time, that points to a fluctuating variance, which can only occur in a nonstationary time series. By contrast, the plot of a stationary time series would show steady fluctuations without a trend. After inspecting the plot of a time series, you can use more formal statistical tests to confirm whether the time series is stationary or nonstationary.

Figure 17-1 shows an example of a stationary time series.

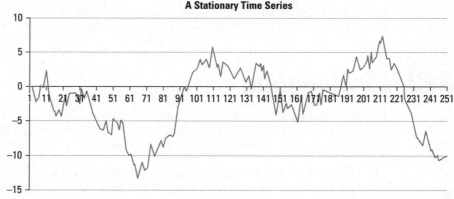

Figure 17-1:
A stationary
time series.

© John Wiley & Sons, Inc.

The time series "randomly" drifts up and down over time — that is, there's no trend. Also, the size of the fluctuations is constant over time, which is another indicator that the time series is stationary.

Figure 17-2 shows an example of a nonstationary time series.

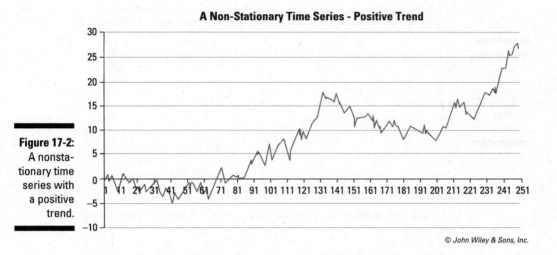

A Non-Stationary Time Series - Positive Trend

© John Wiley & Sons, Inc.

Figure 17-2:
A nonstationary time series with a positive trend.

Although the fluctuations in the time series are stable, there's a pronounced positive (upward) trend. That is an indication that the time series is nonstationary.

Figure 17-3 shows a different example of a nonstationary time series.

Although the time series in Figure 17-3 doesn't have a trend, the size of the fluctuations is increasing over time. That's a sign of nonconstant variances, which is inconsistent with a stationary time series.

There are two types of stationarity: weak form and strong form. ARIMA modeling may be performed if a time series exhibits weak form stationarity. The conditions required for strong form stationarity are unlikely to be met in practice.

When trying to get your head around the statistics of time series analysis, it is important to be clear about what a time series is. From one point of view, a time series is simply a series of observations over time. But in statistics, we treat these observations as the outcomes of a series of random variables over time. Each of these random variables has its own distribution. In simple terms, *stationarity* means that we can reasonably assume that these random variables have basically the same distributions as time progresses.

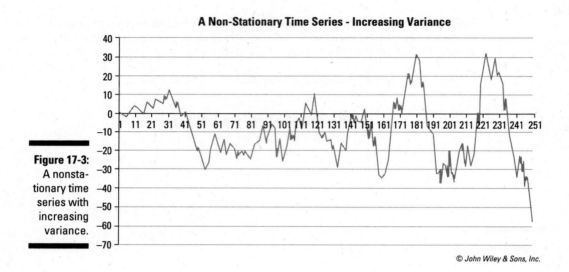

Figure 17-3:
A nonsta-
tionary time
series with
increasing
variance.

Weak form stationarity

A time series exhibits *weak form stationarity* if it meets the following three
conditions:

- ✔ The expected value (mean) of each element of the time series is constant over time.

- ✔ The variance of each element of the time series is constant over time.

- ✔ The value of the covariance of each element in the time series with its
  own past values depends only on the *lag* between the elements (that is,
  the time between the elements). You may recall from Chapter 5 that covariance is a measure of association between two random variables — that
  is, how much their values tend to move together in a pattern. Loosely
  speaking, weak form stationarity requires that the only thing linking the
  behavior between two points in time is how far apart they are.

Autocovariance and autocorrelation

The covariance between two elements of the same time series is known as
autocovariance. For example, suppose the elements of a time series are written like this:

$$x_t, x_{t-1}, x_{t-2}, x_{t-3}, \ldots$$

x_t is the current value of the time series. x_{t-1} is the value of the time series one
period in the past (that is, with lag 1), x_{t-2} is the value of the time series two
periods in the past (that is, with lag 2), and so forth.

With a weakly stationary time series, the covariance between x_{t-2} and x_{t-4} is the same as the covariance between x_{t-3} and x_{t-5}, a lag of two in both cases. Similarly, the covariance between x_{t-3} and x_{t-6} is the same as the covariance between x_{t-4} and x_{t-7}, a lag of three in both cases.

Autocorrelation refers to the correlation between elements of the same time series. Its relationship with autocovariance is analogous to the relationship between correlation and covariance.

Adjustments for nonstationarity

Under some conditions, it may be possible to transform a nonstationary time series into a stationary one. One commonly used approach is known as *first differencing*. You create a new time series in which each element is the *difference* between two consecutive elements in the original time series. The first difference of a time series is expressed this way:

$$z_t = x_{t-1} - x_t$$

where

x_t is an element of the original time series.

x_{t-1} is the value of the time series one period in the past.

z_t is an element of the transformed time series.

For example, using the time series in Figure 17-2, Figure 17-4 is a graph of the time series after first differencing.

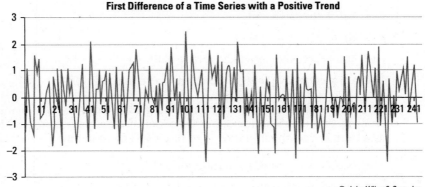

First Difference of a Time Series with a Positive Trend

Figure 17-4:
First difference of a time series with a positive trend.

© John Wiley & Sons, Inc.

The graph in Figure 17-4 shows that the trend has been removed from the data, making it into a stationary time series. The reason this works is that the original trend was a line. One way of defining a line is that it moves up or down by a constant amount over each unit on the horizontal axis. Therefore, by taking distances you remove the slope of the line.

If first differencing isn't sufficient to transform a nonstationary time series into a stationary time series, you can try *second differencing*. In this case, each element in the new time series is the difference between consecutive elements in the first-differenced time series.

Steps used in ARIMA modeling

Once you have determined that a time series is stationary, you can apply the ARIMA modeling process. This entails four key steps:

- ✔ Model identification
- ✔ Parameter estimation
- ✔ Model checking
- ✔ Forecasting

Model identification

In the model identification phase, you determine the properties of a time series; then you choose a structural form for modeling the time series. As mentioned earlier, you are now treating the time series as a series of random variables. This is generally referred to as a *random process*. The basic types of ARIMA models are as follows:

- ✔ Autoregressive (AR)
- ✔ Moving average (MA)
- ✔ Autoregressive moving average (ARMA)
- ✔ Autoregressive integrated moving average (ARIMA)

A time series may be primarily an autoregressive process, a moving average process, or a combination of both. To identify the type of process followed by a time series, the first step is to plot two key functions:

- ✔ Sample autocorrelation function (ACF)
- ✔ Sample partial autocorrelation function (Partial ACF)

The ACF shows the correlations between the elements of a time series as a function of their lags. The partial ACF shows the correlations between the elements of a time series for each lag, *holding constant* the impact of all other lags. You can use the patterns shown in these plots to identify the appropriate type of ARIMA model.

For example, Figure 17-5 shows the ACF for a time series consisting of the price of Microsoft stock from January 1, 2013 to January 1, 2014.

Figure 17-5:
Auto-
correlation
function
(ACF) for
Microsoft
stock
prices.

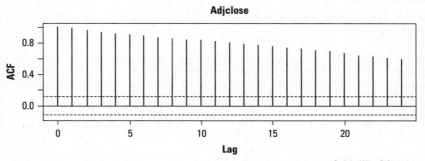

© John Wiley & Sons, Inc.

Each "spike" on the graph is the autocorrelation between elements of the time series with a specified lag. In this case, the spikes are declining very slowly as the lag increases — a sign of a nonstationary time series.

Any spike that rises above the upper dashed line or falls below the lower dashed line is "statistically significant" or significantly different from zero. The placement of the dashed lines is based on a hypothesis test similar to the ones described in Chapters 15 and 16.

Figure 17-6 shows a plot of Microsoft stock prices during the same time period.

Figure 17-6:
Price of
Microsoft
stock
January 1,
2013–
January 1,
2014.

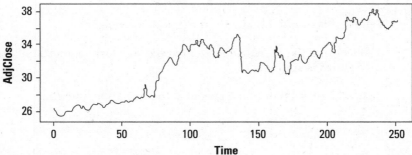

© John Wiley & Sons, Inc.

Figure 17-6 clearly shows an upward trend in the time series, which indicates nonstationarity.

Figure 17-7 illustrates the partial autocorrelation function for the price of Microsoft stock from January 1, 2013 to January 1, 2014.

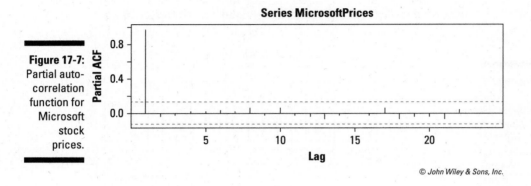

Series MicrosoftPrices

Figure 17-7: Partial auto-correlation function for Microsoft stock prices.

In order to convert the Microsoft prices to a stationary series, first differencing is applied. Figure 17-8 shows the results.

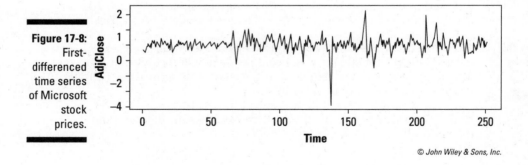

Figure 17-8: First-differenced time series of Microsoft stock prices.

Figure 17-9 shows the ACF for the first-differenced data.

The fact that the spikes no longer exceed the dashed lines indicates that the time series is stationary.

Figure 17-10 is a plot of the partial autocorrelation function.

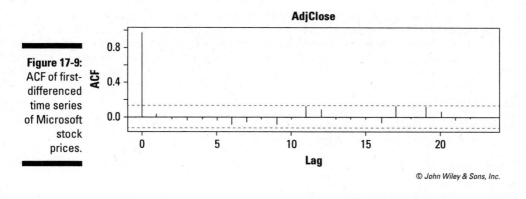

Figure 17-9: ACF of first-differenced time series of Microsoft stock prices.

© John Wiley & Sons, Inc.

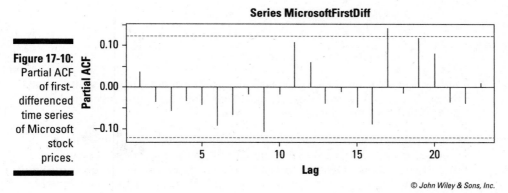

Figure 17-10: Partial ACF of first-differenced time series of Microsoft stock prices.

© John Wiley & Sons, Inc.

The autocorrelation function and partial autocorrelation function for the first-differenced Microsoft stock prices can now be used to identify whether the time series follows a moving average process, an autoregressive process, or some combination of the two.

Moving average (MA) processes

A *moving average* process is composed of a linear combination of *white noise* terms.

A *white noise process* is a special type of time series in which the elements are independent of each other and have no trend. This means that the value of the time series at any given point does not depend in any way on its previous values. Another defining characteristic of a white noise process is that all the random variables that make up the process have the same distribution, and the mean of that distribution is zero. The process evolves randomly over time.

To get a sense of what's going on here, think of a simple random walk. Flip a coin. If it comes up heads take a step forward. If it comes up tails, take a step back. Over time, if the coin is fair, the heads will balance out the tails and you'll end up back where you started. We treat the random fluctuations in stock prices in a similar way, though with a broader distribution of outcomes than just up or down.

In general, a moving average process is written as follows:

$$x_t = \mu + u_t + \theta_1 u_{t-1} + \theta_2 u_{t-2} + \ldots + \theta_q u_{t-q}$$

where

μ = A constant

u_t = A white noise term at time t

θ_t = The coefficient of the white noise term at time t

q = The number of *lags* in the process

This is known as an MA(q) process, because there are q lagged white noise terms. For an MA(q) process, the autocovariance and autocorrelation functions equal zero for all lags greater than q.

Unlike some of the other modeling techniques we've explored (such as regression analysis), the coefficients in ARIMA models are not calculated directly based on a formula. Instead, they are estimated based on comparing a large number of random simulations to the observed data. As an example, a simulated version of the MA(1) process

$$x_t = 0.9 u_{t-1} + u_t$$

is illustrated in Figure 17-11.

Figure 17-12 shows the autocorrelation function for this process.

Figure 17-11: Plot of simulated MA(1) process.

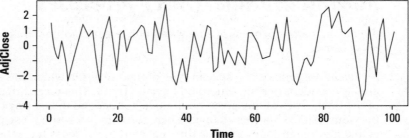

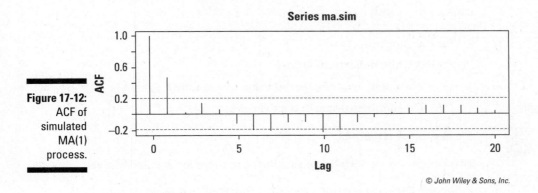

Series ma.sim

Figure 17-12: ACF of simulated MA(1) process.

The graph shows a statistically significant spike at a lag of one; most of the remaining spikes are not statistically significant. This is indicative of an MA(1) process.

Figure 17-13 is a graph of the partial autocorrelation function for this process.

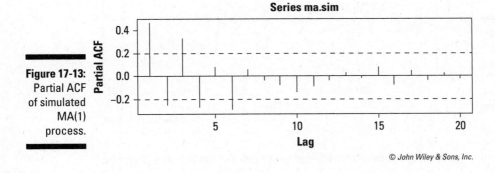

Series ma.sim

Figure 17-13: Partial ACF of simulated MA(1) process.

The graph shows that the partial autocorrelations decline very gradually in magnitude, which is also indicative of an MA(1) process.

Autoregressive (AR) processes

An *autoregressive process* is composed of a *linear combination* of its own past values and a white noise term. In general, this is written as follows:

$$x_t = \mu + \phi_1 x_{t-1} + \phi_2 x_{t-2} + \ldots + \phi_p x_{t-p} + u_t$$

where

μ = A constant

u_t = A white noise term at time t

ϕ_t = The coefficient of a lagged value of x at time t

p = The number of lags in the process

ϕ is the Greek letter *phi*.

This process is an AR(p) process, because there are p lagged terms.

As an example, a simulated version of the AR(1) process

$$x_t = 0.9x_{t-1} + u_t$$

is illustrated in Figure 17-14. Note that unlike the MA(1) process pictured in Figure 17-11, this model uses the lagged value of the stock price combined with a white noise term.

Figure 17-14:
Plot of
simulated
AR(1)
process.

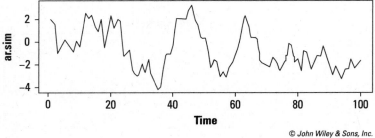

© *John Wiley & Sons, Inc.*

Figure 17-15 shows the autocorrelation function for this process.

The spikes are declining in magnitude very slowly, which is characteristic of an AR process.

Figure 17-16 is a plot of the partial autocorrelation function for this process.

Most of the spikes aren't statistically significant, which is characteristic of an AR function.

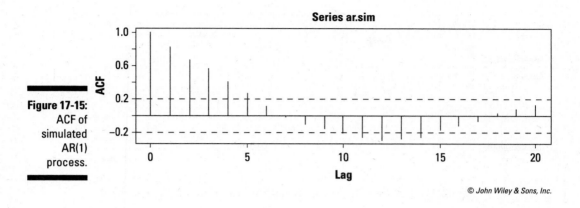

Figure 17-15:
ACF of
simulated
AR(1)
process.

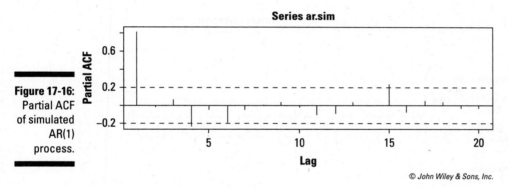

Figure 17-16:
Partial ACF
of simulated
AR(1)
process.

Autoregressive moving average (ARMA) processes

An *autoregressive moving average process* is a combination of an autoregressive and a moving average process. In general, this is written as follows:

$$x_t = \mu + \phi_1 x_{t-1} + \phi_2 x_{t-2} + \ldots + \phi_p x_{t-p} + u_t + \theta_1 u_{t-1} + \theta_2 u_{t-2} + \ldots + \theta_q u_{t-q}$$

This is known as an ARMA(p,q) process, because there are p lagged terms and q white noise terms.

For example, here's a simulated version of the ARMA(1,1) process:

$$x_t = 0.6 x_{t-1} + 0.8 u_{t-1} + u_t$$

Figure 17-17 illustrates the results.

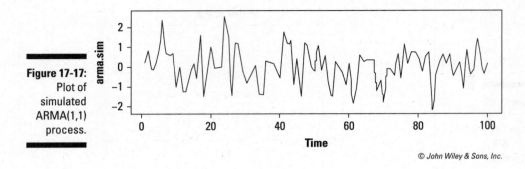

Figure 17-17:
Plot of
simulated
ARMA(1,1)
process.

© John Wiley & Sons, Inc.

Figure 17-18 shows the autocorrelation function of this process.

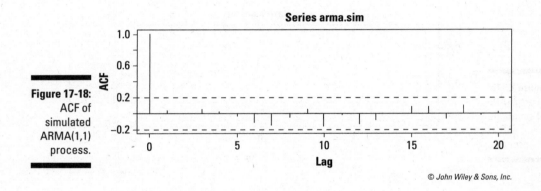

Figure 17-18:
ACF of
simulated
ARMA(1,1)
process.

© John Wiley & Sons, Inc.

The spikes are declining in magnitude very slowly, which is characteristic of an ARMA process.

Figure 17-19 displays the partial autocorrelation function of this process.

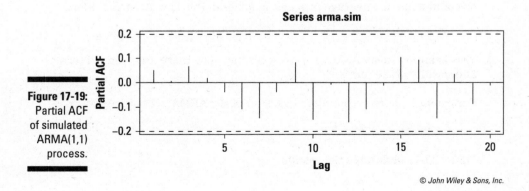

Figure 17-19:
Partial ACF
of simulated
ARMA(1,1)
process.

© John Wiley & Sons, Inc.

The spikes are declining in magnitude very slowly, which is characteristic of an ARMA process.

Autoregressive integrated moving average (ARIMA) processes

An autoregressive integrated moving average (ARIMA) process is an ARMA process in which the original time series has been differenced in order to make it stationary. For example, an ARMA(2, 2) model that has been first-differenced is known as an ARIMA(2, 1, 2) model. In general, an ARIMA model is written as ARIMA(p, d, q), where p is the number of AR terms, q is the number of MA terms, and d is the number of times that the time series has been differenced.

Parameter estimation

Once you have tentatively identified the form of an ARIMA model, the next step is to estimate the coefficients or parameters of the model. You can accomplish this with a couple of advanced techniques:

- Regression analysis
- Maximum likelihood estimation

Model checking

Once you've estimated a model, you can determine the quality of the results with a measure known as an *information criterion*. This is a test statistic whose value is minimized for the best fitting model. One of the most widely used information criterion is known as Akaike's Information Criterion (AIC).

As an example, you can model the first-differenced data for Microsoft stock prices with several different forms; each is compared based on its AIC. Table 17-1 shows the results.

Each row in the table shows the number of lagged terms (p), the number of white noise terms (q), and the number of times the series was differenced (d). In this case, the Microsoft stock price time series is nonstationary; the first difference of this time series is stationary.

The first row shows the results of estimating an AR(1) model (because $p = 1$ and $q = 0$). The second row shows the results of estimating an AR(2) model (because $p = 2$ and $q = 0$). The third row shows the results of estimating an MA(1) model (because $p = 0$ and $q = 1$). The last row shows the results of estimating an ARIMA(2, 1, 2) model (because $p = 2$, $d = 1$, and $q = 2$). Since this

Table 17-1	Results of Fitting Microsoft Stock Prices		
p	*d*	*q*	*AIC*
1	1	0	377.10
2	1	0	378.92
0	1	1	377.08
0	1	2	378.96
1	1	1	379.04
2	1	1	379.21
1	1	2	379.37
2	1	2	374.74

model has the *smallest* value for the AIC statistic, this is the best choice of a model for the Microsoft stock time series. The estimated model is:

$$z_t = 1.8512z_{t-1} - 0.9827z_{t-2} - 1.8805u_{t-1} + 0.9999u_{t-2} + u_t$$

where

z_t is the first difference of the original time series at time *t*

u_t is a white noise term at time *t*

We obtained the AIC values and estimated parameters using the R programming language with the function ARIMA. In general, developing ARIMA models is dependent on having a statistical package at your disposal. In addition to R, SAS and SPSS are both widely used in business applications.

Forecasting

Once an ARIMA model has been estimated, you can forecast future values for the time series. The predicted value for Microsoft stock on January 2, 2014, is 37.37.

We obtained this prediction using the R programming language with the function PREDICT.

Simulation Techniques

Simulation techniques are used to model the behavior of a time series by making assumptions about its statistical behavior. This approach has the advantage that it doesn't require large amounts of historical data to be implemented, as ARIMA modeling does. The main drawback to simulation techniques is that they can be extremely time-consuming to implement.

This section covers two types of simulation techniques:

- ✔ Historical simulation
- ✔ Monte Carlo simulation

Historical simulation

Historical simulation is a technique you can use to generate a probability distribution for a variable's future value based on a time series of its own past values. The success of this technique depends heavily on the assumption that the past will be repeated in the future.

The historical simulation approach has one key advantage over other forecasting techniques: It doesn't require any specific assumptions about the following:

- ✔ The probability distribution followed by the elements of a time series
- ✔ The statistical relationship between the elements of a time series (that is, the correlation and covariance)
- ✔ The moments of the elements of a time series (mean, variance, and so forth)

In many financial applications, the elements of a time series are assumed to be normally distributed. Further, the elements of a time series are frequently assumed to be *independent* of each other — in other words, they do not influence each other.

If variables are independent of each other, they will have a correlation (or covariance) of zero between them.

The historical simulation approach is useful when it's difficult to justify making these types of assumptions or if the statistical properties of the time series are unknown.

The historical simulation approach has some *disadvantages* compared with other forecasting techniques:

- ✔ Its accuracy depends on the extent to which the past behavior of a time series continues to be replicated in the future.

> ✔ The results depend on how much past data is used; for example, the results are likely to be different if 30 days of data are used compared to 2 years of data.
>
> ✔ The technique can be extremely time-consuming compared with other forecasting techniques.

Application: Using historical simulation to compute Value at Risk (VaR)

Value at Risk (VaR) is a measure used to determine how much risk an investor is exposed to by holding an asset or a portfolio of assets. A probability distribution is derived for the return on an asset or portfolio; the left tail of this distribution represents the worst outcomes that may occur with a given probability.

For example, suppose that as of August 25, 2014, an investor holds a portfolio of stocks that is worth $1,000,000. The returns to the portfolio over the previous ten trading days are given in Table 17-2.

Table 17-2	Time Series of Portfolio Returns
Day	**Return %**
August 11, 2014	1.21
August 12, 2014	0.74
August 13, 2014	−0.43
August 14, 2014	0.02
August 15, 2014	−0.26
August 18, 2014	0.63
August 19, 2014	0.31
August 20, 2014	−0.78
August 21, 2014	0.54
August 22, 2014	0.12

You can use historical simulation to create a probability distribution for these returns, which can then be used to determine the VaR of this portfolio. Assume that the investor wants to compute VaR at the 90 percent confidence level. There is a 10% chance that the investor will lose this amount (VaR) or more on the following trading day.

The first step is to sort the historical returns from the worst to the best; these are $a - b$.

–0.78 percent, –0.43 percent, –0.26 percent, 0.02 percent, 0.12 percent, 0.31 percent, 0.54 percent, 0.63 percent, 0.74 percent, 1.21 percent

Figure 17-20 illustrates these returns in a histogram.

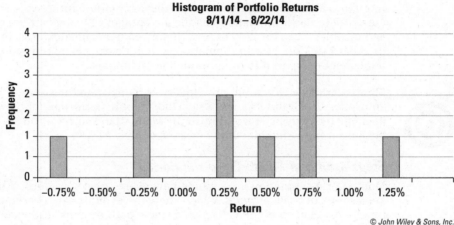

Figure 17-20:
Histogram
of portfolio
returns.

© John Wiley & Sons, Inc.

A *histogram* is a graph you can use to represent a discrete probability distribution. Each bar represents a single category, which is a value or range of values. The heights of the bars represent the frequency of each category.

On the histogram in Figure 17-20, the categories are ranges of returns. For example, the bar labeled –0.75% includes all returns of at least –1.00% up to (but not including) –0.75%. The bar labeled –0.50% includes all returns of at least –0.50 percent up to (but not including) –0.25 percent.

With only ten returns in the sample, the Value at Risk with 90 percent confidence is the 10th worst outcome, which is a loss of 0.78 percent. Since the portfolio is worth $1,000,000, a loss of 0.78 percent translates into a dollar loss of $(0.0078)(1,000,000) = \$7,800$.

This can be interpreted as follows: Over the coming trading day, there is a 10 percent chance that the losses to the portfolio will exceed $7,800. Another way of looking at this is that over the coming trading day, there is a 10 percent chance that the value of the portfolio will fall below $1,000,000 - \$7,800 = \$992,200$.

Note that the mean of this distribution is a gain of 0.21 percent. You obtain this by averaging all ten returns. You can use this result as a forecast of the actual return over the coming trading day.

Monte Carlo simulation

Monte Carlo simulation is a technique in which random numbers are substituted into a mathematical model in order to compute future values of a variable. Monte Carlo simulation has the advantages of being extremely powerful and flexible; it can be used to price highly complex financial assets. The main drawback to Monte Carlo simulation is that it's extremely time-consuming to implement compared to other modeling techniques.

 Monte Carlo is a city in the principality of Monaco, in the south of France; it's known for the Grand Prix and its casinos. Monte Carlo simulation was given its name by a physicist working on the Manhattan Project during World War II, because it reminded him of the casinos in Monte Carlo.

Steps used in a Monte Carlo simulation

Monte Carlo simulation is conducted in several stages or steps. You generate a series of trials or *paths,* which show the evolution of a variable over time. You get each of the values on a single path by substituting a random number into a statistical model. At the end of each path, you obtain a forecasted value. The values are averaged to get a single forecasted value for the variable.

Simulating stock prices with a Monte Carlo simulation

For the specific case of simulating stock prices, you take the following steps:

1. **Specify a statistical model that explains how the stock price evolves over time.**

2. **Generate random numbers.**

3. **Substitute the random numbers into the statistical model.**

4. **Repeat this process multiple times until an entire path is obtained that shows the evolution of the stock price over a specified time horizon. Each path provides a simulated value for the stock price at a specified point in the future.**

5. **Generate a large number of paths; the average final value over all the paths is used as a forecast of the stock price.**

Specifying a statistical model

Simulating a stock's price with Monte Carlo simulation requires several key assumptions:

- ✔ The statistical process followed by the stock over time
- ✔ The probability distribution that the stock's price changes follow over time
- ✔ The moments of the stock's price

You can model the statistical process that a stock's price follows over time with a *stochastic differential equation* (SDE). A stochastic differential equation shows how the value of an asset changes over time. It's based on several assumptions, such as the probability distribution followed by the asset price, how the volatility of the asset price changes over time, and so forth.

Geometric Brownian motion

One popular choice for modeling stock prices is a stochastic differential equation known as geometric Brownian motion (GBM).

Geometric Brownian motion is the process used to model stock prices by the Black-Scholes option pricing model.

The geometric Brownian motion process is given by the following stochastic differential equation:

$$dS = \mu S dt + \sigma S dZ$$

where

S is the price of the stock.

dS is an instantaneous change in the stock price.

μ is the drift rate of the asset (how quickly S grows over time).

dt is an instantaneous change in time.

σ is the volatility of the stock price.

dZ is another stochastic process known as *Brownian motion*

Volatility is another name for the standard deviation of the continuously compounded return to the stock.

Brownian motion is a continuous process that randomly drifts up and down in such a way that its average value is zero and its variance increases over time.

Figure 17-21 is an illustration of Brownian motion. The figure shows that Brownian motion aimlessly wanders around a value of zero, but can move further and further away from zero as time elapses.

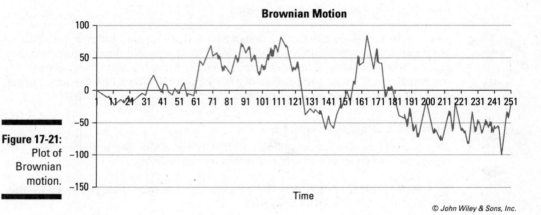

Figure 17-21:
Plot of
Brownian
motion.

© John Wiley & Sons, Inc.

The Monte Carlo process isn't continuous but is instead implemented as a series of discrete time steps along a path. For pricing stocks, the GBM process can be implemented as follows:

$$\Delta S = \mu S \Delta t + \sigma S \varepsilon \sqrt{\Delta t}$$

where

- ΔS is a small change in the stock price.

- Δt is a small change in time.

- ε is a random number chosen from the standard normal distribution.

ε is the Greek letter *epsilon*.

There are infinitely many normal distributions, each uniquely characterized by its mean and standard deviation. The standard normal distribution has a mean of 0 and a standard deviation of 1.

Generating random numbers

One of the biggest challenges of correctly implementing Monte Carlo simulation is correctly generating random numbers. You choose a random number from a specified probability distribution. For example, you could randomly choose a number from the set {1, 2, 3, 4, 5} in such a way that each number is equally likely to be chosen. As another example, you could choose a number from the interval 0 to 1 in such a way that a number from each interval is equally likely to be chosen. In this case, a number between 0.1 and 0.2 is equally likely to be chosen as a number between 0.2 and 0.3, a number between 0.3 and 0.4, and so forth. This is because each of these intervals has the same width (0.1). As another example, you could choose a number from the normal distribution.

Excel has a built-in random number generator which you can access with the function RAND. This chooses a number between 0 and 1 in such a way that equal-sized intervals between 0 and 1 are equally likely to be chosen. To generate a random number, enter the formula =RAND(). (Notice that you don't enter a value between the parentheses.) The Data Analysis Toolpak also contains a large number of different random number generators.

Excel doesn't have a random number generator for the standard normal distribution, but there's a way to work around this. The function =NORMSINV (RAND()) chooses a random number between 0 and 1 and then transforms it into a standard normal random number with the NORMSINV command (NORMSINV stands for the inverse of the standard normal distribution).

For example, suppose a researcher wants to create a probability distribution for the price of ABC stock one year in the future. The stock is currently selling for $100, the annual drift of the stock's price is 5 percent, and the annual volatility is 30 percent. The researcher wants to use Monte Carlo simulation to simulate ABC's price using a series of paths, each consisting of 12 one-month time steps. In this case, Δt equals $1 \div 12 = 0.08333$.

Table 17-3 shows the calculations needed for a single path.

The stock price is initially $100. For each time step, a random number is chosen from the standard normal distribution. The result appears in the ε column. This is substituted into the discrete version of geometric Brownian motion to obtain a simulated change in the stock price over the next time step (the next month). This change is added to the current value of S to obtain the simulated value in the next time step. For example, the change in S over the first month is 20.28, so the simulated stock price in one month is therefore $100 + 20.28 = 120.28$.

Table 17-3		Time Series of Portfolio Returns	
Time	**S**	**ε**	**$\Delta S = \mu S \Delta t + \sigma S \varepsilon \sqrt{\Delta t}$**
0	100.00	2.2941	20.28
1	120.28	0.2149	2.74
2	123.02	−0.8880	−8.95
3	114.08	1.8354	18.61
4	132.68	−0.5759	−6.07
5	126.62	−0.5869	−5.91
6	120.71	−1.8972	−19.33
7	101.38	0.0813	1.14
8	102.52	0.8101	7.62
9	110.13	1.6526	16.22
10	126.36	−0.3860	−3.70
11	122.66	−0.1299	−0.87
12	121.79	0.5710	6.53

The stock price in one year is estimated to be $121.79. You can think of this as a forecasted value for the stock price in one year.

The standard error of an estimated value and confidence intervals

One of the drawbacks to the Monte Carlo method is that the value simulated on each path is unlikely to be a good estimate of the actual future value of the stock price. Normally, you would generate a very large number of paths. Then, you average the future values of the stock to get a single forecasted price. This leaves one important question unanswered: How much uncertainty is there associated with this estimate?

You can determine the uncertainty with a measure known as the *standard error*. You use the standard error to construct a confidence interval to provide a measure of how much uncertainty is associated with a simulated price.

A *confidence interval* is a range of values that is likely to contain the true value of an estimated variable. For example, you can construct a confidence interval for the mean of a population. A confidence interval is associated with a confidence level, such as 95 percent, that indicates how much uncertainty is associated with the confidence interval.

The standard error is computed as follows:

$$\frac{s}{\sqrt{n}}$$

where

    s is the standard deviation of the stock price across the simulated paths.

    n is the number of paths generated.

The standard deviation of a sample of data is computed like so:

$$s = \sqrt{\frac{\sum_{i=1}^{n}(x_i - \bar{x})^2}{n-1}}$$

where

    x_i is a single element in the sample.

    $\bar{x}$ is the mean of the elements in the sample.

    n is the sample size.

You can then construct a 95 percent confidence interval:

    Lower limit = Estimated price $-1.96\times$ Standard error

    Upper limit = Estimated price $+1.96\times$ Standard error

The value 1.96 is used in a 95 percent confidence interval because 95 percent of the standard normal distribution falls within 1.96 standard deviations of the mean.

As an example, suppose that the researcher in the previous example decides to generate 100,000 paths in order to estimate the correct price of ABC stock. Figure 17-22 shows a histogram that summarizes the results.

The mean price of these trials is $108.74. This is the estimated (forecasted) price of ABC stock. If the standard deviation is 7.49, then the standard error of the estimated price would be this:

$$\frac{s}{\sqrt{n}} = \frac{7.49}{\sqrt{100,000}} = 0.0236$$

The standard error of the estimated price is $0.0236. A 95 percent confidence interval for the estimated price would then be this:

    Lower limit = $108.74 - 1.96 \times 0.0236 = 108.69$

    Upper limit = $108.74 + 1.96 \times 0.0236 = 108.79$

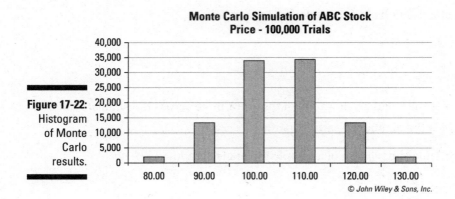

Figure 17-22: Histogram of Monte Carlo results.

This can be interpreted as follows: If this simulation is repeated 100 times, the interval $108.69 to $108.79 will contain the true price of ABC stock in 95 of these cases.

Simulating Value at Risk with Monte Carlo simulation

The paths generated by Monte Carlo simulation generate a probability distribution for the value of an asset. You can compute the VaR of an asset or a portfolio by identifying the left tail of the distribution, similar to historical simulation.

Referring to the example of the ABC stock, suppose the researcher wants to determine the VaR of ABC stock at the 95 percent level of confidence. With 100,000 paths, the 5,000 worst outcomes would be the threshold of the left 5 percent tail of the distribution. Suppose that the 5,000th worst outcome is a price of $94.77. Then the Value at Risk for ABC stock is a loss of $100 – $94.77 = $5.23 per share. For an investor who holds 100 shares of ABC stock, there is a 5 percent chance that his or her losses over the coming trading day will exceed 100($5.23), or $523.00.

Once again, we've been through a lot of statistical methods in this chapter. But as always, remember that much of the heavy lifting involved in applying these statistical techniques is done by software. In the business world, this software is very easy to use as well as quite powerful. Unfortunately, it is also quite expensive. Chapter 18 takes you through some of the technical details involved in applying statistical processes using software that is either free or that you probably already have on your computer.

Chapter 18

Crunching Numbers: Performing Statistical Analysis on Your Computer

*P*erforming any type of statistical analysis usually requires the use of a specialized program. There are many available choices, ranging from the very basic to the extremely sophisticated. Some cost thousands of dollars — others you can download for free from the Internet.

This chapter introduces the statistical capabilities of three widely used such programs:

- ✔ Excel
- ✔ Visual Basic for Applications (VBA)
- ✔ The R programming language and interface

This chapter does assume some familiarity with coding in the Microsoft Office environment, particularly with VBA. If that environment is totally unfamiliar to you, you may want to bone up on some of the basics and then come back to this chapter. One good resource is *Excel VBA Programming For Dummies* by John Walkenbach (Wiley, 2013).

Excelling at Excel

Microsoft Excel is a hugely popular spreadsheet program that offers a wide range of capabilities. One of its great strengths is its ability to perform complex statistical analysis with built-in commands. Excel is also very useful for organizing and storing data. Many statistical packages are capable of reading data from Excel files and can also export their own results as Excel files. In addition to its built-in statistical functions, the capabilities of Excel have been extended with an add-in package known as the Analysis ToolPak.

 The Analysis ToolPak is no longer included with the Mac version of Excel. (You can download a free version of StatPlus, which has similar functionality, from `http://statplus-mac.en.softonic.com/mac/download`. It is produced by the same company, Softonic, that makes the Analysis ToolPak.)

Key Excel statistical functions

Excel contains a numerous built-in statistical functions, including the following:

✔ Summary measures, such as mean, variance, and standard deviation

✔ Measures of association, such as correlation and covariance

✔ Functions for computing probabilities for discrete distributions, such as the binomial and Poisson

✔ Functions for computing probabilities for continuous distributions such as the normal, Student's t-distribution, and chi-square

✔ Regression analysis functions, such as slope and intercept

In each case, the function requires one or more inputs, usually in a specific order. You can enter these directly or use cell references instead. As a simple example, suppose the average gas price in New York City is $3.59, and is stored in cell A1. The average gas price in New Jersey is $3.29, stored in cell A2. To determine the greater of the two, you use the function MAX as follows:

= MAX(3.59, 3.29)

To implement an Excel function, you enter an equals sign, followed by the name of the function, a left parenthesis, the inputs to the function, and a right parenthesis. Then press the Enter key. After you type the equals sign, the name of the function, and the left parenthesis, Excel provides a "ribbon" that lists all required inputs. In this case, the ribbon shows the following information:

= MAX(number 1, [number 2],...)

This shows that MAX requires at least one numerical input. *number 2* is surrounded by square brackets, which indicates that it's optional. The three dots indicate that you can supply as many numbers as you like.

Alternatively, you can implement the function as follows:

= MAX(A1, A2)

In this case, you use the cell references A1 and A2 instead of directly entering the two values. The advantage of this approach is that references to these cells will change automatically if you change the values in cells A1 and A2.

In both cases, the function returns the correct value of 3.59.

Another way to implement a function is by clicking the *fx* icon to the left of the formula bar. In the example in Figure 18-1, the *fx* icon appears directly above column D.

Figure 18-1:
The *fx* icon
in Excel.

Source: Alan Anderson

After clicking on the *fx* icon, a window containing all of Excel's functions opens. To find the MAX function, you enter **MAX** in the "Search for a function" field; the next step is to click the Go button on the right. (If you misspell the function name, Excel may not be able to find it. Just try again!) See Figure 18-2.

Once you find the MAX function, click the OK button. This produces a window that prompts you for each required value. The window also contains a brief explanation of the function and its required inputs, as shown in Figure 18-3.

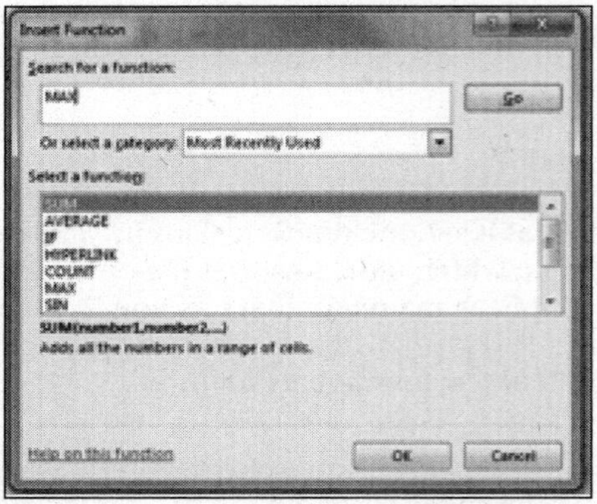

Figure 18-2:
The Insert
Function
window in
Excel.

Source: Alan Anderson

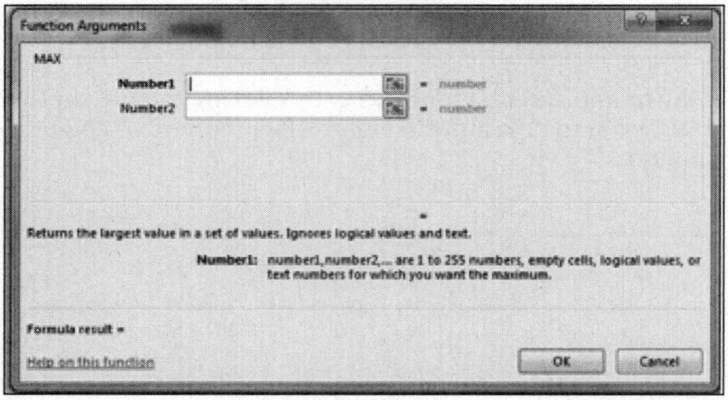

Figure 18-3:
The
Function
Arguments
window.

Source: Alan Anderson

The examples in this section show how to compute summary measures for a sample of stock returns and how to determine the likelihood of default for the bonds in a portfolio, using Excel's built-in functions.

Computing summary measures in Excel

Table 18-1 shows the steps used to compute summary measures for a sample of returns to a stock. The data are entered in two columns; column A contains the year in which each return occurred, and column B contains the actual return for each year. Column labels are in row 1, and the years and returns are found in rows 2 through 6. The appropriate Excel functions are shown in cells B7 through B9.

Table 18-1   Computing Summary Measures for a Sample in Excel

	A	B
1	Year	Return
2	2011	3.40 percent
3	2012	2.70 percent
4	2013	8.10 percent
5	2014	5.90 percent
6	2015	6.40 percent
7	MEAN	= AVERAGE(B2:B6)
8	VARIANCE	= VAR.S(B2:B6)
9	STANDARD DEVIATION	= STDEV.S(B2:B6)

Table 18-1 shows that in order to compute the mean of a sample in Excel, you use the function AVERAGE (no distinction is made between a sample and a population when using the AVERAGE function). For computing the sample variance, you use VAR.S; for a sample standard deviation, you use STDEV.S.

Unlike the mean, the formulas for computing a population variance and a population standard deviation are slightly different from the formulas for computing a sample variance and a sample standard deviation. The functions that provide population variance and population standard deviation are VAR.P and STDEV.P, respectively.

In this case, the results are as follows:

✔ Sample mean = 0.053 = 5.3 percent

✔ Sample variance = 0.00049

✔ Sample standard deviation = 0.02224 = 2.224 percent

Because the sample data consist of returns, which are measured as a percentage, the mean and standard deviation are also measured as percentages. The variance is actually measured as a percentage *squared.* This is difficult to interpret, so standard deviation is more likely than variance to be used as a measure of uncertainty.

Computing probabilities in Excel

Suppose a portfolio manager is responsible for the performance of an extremely risky bond fund that contains ten "junk" bonds. Each bond comes from a different sector of the economy, so the performance of each bond can be treated as independent from the other bonds in the fund. Suppose the portfolio manager has determined from historical experience that each bond has a 40 percent chance of defaulting over the coming year.

The portfolio manager determines that the probability distribution which can best explain the default behavior of the bonds is the *binomial distribution*. (Chapter 4 introduces this distribution and discusses it at length.) Based on this assumption, the portfolio manager can use Excel to create a probability distribution for the number of bonds that default over the coming year.

The binomial distribution is used when independent "trials" are taking place, and each trial can have only one of two possible outcomes. Here, the performance of each bond can be thought of as a trial where only one of two things can happen: default or no default.

Table 18-2 shows the steps that would be used to compute binomial probabilities for the portfolio manager.

Table 18-2	**Using the Binomial Distribution Function in Excel**			
	A	**B**	**C**	**D**
1	n	10		
2	p	0.4		
3				
4	Default	Probability	Default	Probability
5	0	=BINOM.DIST (A5, B1, B2, 0)	6	=BINOM.DIST (C5, B1, B2, 0)
6	1	=BINOM.DIST (A6, B1, B2, 0)	7	=BINOM.DIST (C6, B1, B2, 0)
7	2	=BINOM.DIST (A7, B1, B2, 0)	8	=BINOM.DIST (C7, B1, B2, 0)
8	3	=BINOM.DIST (A8, B1, B2, 0)	9	=BINOM.DIST (C8, B1, B2, 0)
9	4	=BINOM.DIST (A9, B1, B2, 0)	10	=BINOM.DIST (C9, B1, B2, 0)
10	5	=BINOM.DIST (A10, B1, B2, 0)		

The Excel function BINOM.DIST is used to compute binomial probabilities. When entering the function BINOM.DIST, the following ribbon appears:

= BINOM.DIST(number_s, trials, probability_s, cumulative)

The variables listed in parentheses show the inputs that are required as well as the order in which they must be entered:

- number_s
- trials
- probability_s
- cumulative

number_s refers to the number of "successes" that occur. In this example, each default is referred to as a *success,* and each non-default is referred to as a *failure.* For example, to compute the probability that no bonds default, number_s is set equal to 0. Similarly, to compute the probability that exactly one bond defaults, number_s is set equal to 1. The remaining probabilities are computed in the same way.

trials refers to the number of times the process is repeated; in this example, each bond in the portfolio is considered to be a trial, so trials is set equal to 10. *probability_s* is the probability of a success on a single trial, which in this case is 0.40, or 40 percent.

cumulative is a binary variable that accepts only one of two values:

- 0 represents false.
- 1 represents true.

The cumulative variable is set to 1 only when computing "cumulative" probabilities. For example, the probability that the number of defaults is *less than or equal to* 5. Here, the portfolio manager is interested in the probabilities of specific numbers of defaults, not cumulative probabilities, so cumulative is set equal to 0.

Notice that the number of trials and the probability of success are constants; they are stored at the top of the spreadsheet. The number of trials is in cell B1, whereas the probability of success is in cell B2. When you enter the number of trials into the BINOM.DIST formula, use either the value 10 or the cell reference B1. Similarly, when you enter the probability of success into the BINOM.DIST formula, use either the value 0.40 or the cell reference B2. In general, it's safer to use cell references because these values are easy to change.

Notice that the cell references B1 and B2 contain dollar signs ($) in the BINOM.DIST formula. The dollar signs lock these references into place and prevent them from shifting should you copy the formula into other cells. The cells references A5, A6, A7, and so forth are allowed to change because these contain the number of trials, which will be different for each probability you're computing.

Table 18-3 shows the actual probabilities for each possible number of defaults. You can think of this table as a *probability distribution*.

Table 18-3	Computing Default Probabilities in Excel		
Defaults	*Probability*	*Defaults*	*Probability*
0	0.0060	6	0.1115
1	0.0403	7	0.0425
2	0.1209	8	0.0106
3	0.2150	9	0.0016
4	0.2508	10	0.0001
5	0.2007		

The portfolio manager can now see at a glance that the most likely outcome is four defaults, and that the likelihood of no defaults is only 0.0060. This information can also be used to determine the average number of defaults that are likely to occur, the standard deviation of the number of defaults (that is, the uncertainty associated with the number of defaults), and so on.

Updated statistical functions

Microsoft updated many of its built-in functions with the 2010 version of Excel for Windows and the 2011 version of Excel for the Mac. Here are a few of the most important changes made:

- ✔ Increased speed of calculations
- ✔ More consistent naming of functions
- ✔ Introduction of a few completely new functions

In some cases, the algorithms used to implement the functions have been replaced or improved to increase the speed and accuracy of the calculations. For example, according to Microsoft, the BETADIST function was not implemented accurately in earlier versions of Excel, and this has been corrected. (BETADIST is used to compute probabilities for the beta distribution.)

Another major change took place with the spelling of the statistical functions. Many inconsistencies have been eliminated. For example, in older versions of Excel, the population variance was computed with the function VARP, whereas the sample variance was computed with the function VAR. These functions are now spelled VAR.P and VAR.S, respectively. (P stands for *population,* and S stands for *standard deviation.*) Similarly, the population standard deviation was previously spelled STDEVP and the sample standard deviation was spelled STDEV. These have been replaced with STDEV.P and STDEV.S, respectively. All sample measures now end with .S, and all population measures now end with .P.

In spite of all of these changes, Excel has maintained *backward compatibility*. This means that the original spellings of Excel functions still work in the newest versions of Excel.

In addition to the improvements to existing functions, several completely new functions have been added. Table 18-4 shows some of the new functions.

Table 18-4	New Excel Functions
CHISQ.DIST	T.DIST
CHISQ.INV	T.INV
F.DIST	CONFIDENCE.T
F.INV	CONFIDENCE.NORM

CHISQ.DIST is used to compute probabilities under the chi-square distribution, and CHISQ.INV is used to determine the location under the chi-square distribution that corresponds to a given probability. Similarly, F.DIST and T.DIST are used to compute probabilities for the F and Student's t-distribution, respectively. F.INV and T.INV are used to find locations under the F- and T-distributions, respectively.

You can use CONFIDENCE.T and CONFIDENCE.NORM to construct *confidence intervals.* CONFIDENCE.T is based on the Student's t-distribution, which is used when constructing a confidence interval for the population mean with an unknown population standard deviation. CONFIDENCE.NORM is based on the normal distribution, which is used when constructing a confidence interval for the population mean with a known population standard deviation. In both cases, the function returns the *margin of error.* You must compute the actual confidence interval.

For example, you can construct a confidence interval for a population mean as follows:

sample mean ± margin of error

The sample mean is an estimate of the population mean. The margin of error indicates how much uncertainty is associated with this estimate.

For example, suppose a university has 10,000 students with normally distributed grade point averages (GPAs). The university would like to estimate the average GPA of all the students. Due to time constraints, the university will not calculate the actual mean (the population mean) for the entire university. Instead, an analyst will construct a confidence interval from sample data.

The analyst chooses a sample of ten students, among whom the sample mean GPA is 3.10, and the sample standard deviation is 0.25. The university would like to use this information to construct a 95 percent confidence interval for the *population* mean GPA (in other words, for the mean GPA of the entire student body).

Because the analyst knows only the sample standard deviation, he or she uses the Student's t-distribution (discussed in Chapter 13) to construct the confidence interval. The margin of error is based on the Student's t-distribution, and is computed with the following formula:

$$t_{\alpha/2}^{n-1} \frac{s}{\sqrt{n}}$$

where

n is the sample size.

α is the *level of significance.*

s is the sample standard deviation.

$t_{\alpha/2}^{n-1}$ is a location under the Student's t-distribution.

In this example, n is 10, because there are ten GPAs in the sample. The level of significance (α) is 0.05, or 5 percent. This is because the confidence interval is being constructed with a 95 percent level of confidence, and the level of significance equals 1 minus the level of confidence. $s = 0.25$, because this is the standard deviation of the sample GPAs.

The appropriate value chosen from the Student's t-distribution is based on the level of significance and the number of *degrees of freedom,* which equals the sample size minus 1. In this case, the degrees of freedom equals 9.

Using this information, the analyst can compute the Student's t value using Excel's T.INV function:

= T.INV(probability, deg_freedom)

The probability is set equal to $1 - (\alpha/2) = 1 - (0.05/2) = 1 - 0.025 = 0.975$. The degrees of freedom (deg_freedom) equals $n-1(9)$. Therefore, the function with values

= T.INV(0.975,9)

is equal to approximately 2.262. As a result, the margin of error equals the following:

$$t_{\alpha/2}^{n-1} \frac{s}{\sqrt{n}} = 2.262 \left(\frac{0.25}{\sqrt{9}} \right) = 0.189$$

The confidence interval is therefore

$$\bar{x} \pm t_{\alpha/2}^{n-1} \frac{s}{\sqrt{n}} = 3.10 \pm 0.189$$

$\bar{x}$ represents the sample mean in this formula.

The confidence interval can be written as $(3.10 - 0.189,\ 3.10 + 0.189)$ or (2.911, 3.289). This interval provides a sense of how close the population GPA is likely to be to the sample mean of 3.10.

Analysis ToolPak

The Analysis ToolPak is an add-in package included with Excel for Windows (StatPlus for Macs has similar functionality). You can use it to perform several types of statistical procedures:

✔ Hypothesis testing for one or two populations

✔ Regression analysis

✔ ANOVA (analysis of variance)

✔ Generating summary statistics, such as the mean, median, mode, variance, standard deviation, and so on

✔ Computing covariances and correlations between datasets

✔ Generating random numbers

You access the Analysis ToolPak through the Data tab. It appears in the Analysis section on the far right of the ribbon. If the icon doesn't appear on the ribbon, you can install it using the following steps:

1. **Click the File tab.**

2. **Choose Options.**

3. **In the Excel Options window, choose Add-Ins.**

4. **Click the Go button.**

5. **In the Add-Ins dialog box, select Analysis ToolPak.**

6. **Click the OK button.**

Figure 18-4 shows the Data toolbar with the Analysis tab on the far right.

Figure 18-4:
The Data toolbar with Analysis tab.

Source: Alan Anderson

Clicking the Data Analysis icon opens the Data Analysis window. You use a function by selecting it and clicking the OK button. For each function, a dialog box appears that prompts you for the required information.

Figure 18-5 shows the Data Analysis dialog box.

Figure 18-5:
The Data Analysis window.

Source: Alan Anderson

For example, suppose you're conducting a Monte Carlo simulation study that requires 100 numbers that are randomly chosen from the standard normal distribution. These numbers will be used to produce a probability distribution of 100 possible values for the returns to a stock.

The normal distribution (sometimes known as the *bell-shaped curve*) is uniquely characterized by two values: the mean and the standard deviation. There are infinitely many normal distributions. The special case in which the mean equals 0 and the standard deviation equals 1 is known as the *standard normal distribution*. Many applications in a wide variety of disciplines use this distribution.

Excel can generate a random number with the function =RAND(). This number is based on the standard uniform distribution, so a number between 0 and 1 will be generated. The probability of getting a number within a specific range depends only on the width of the range. For example, the probability of getting a number between 0.1 and 0.2 is the same as the probability of getting a number between 0.3 and 0.4, and the same as the probability of getting a number between 0.5 and 0.6, and so forth. In each case, the probability is 0.1 since each of these intervals represents one-tenth of the entire range of possibilities from 0 to 1.

If you need another probability distribution, you must use a different approach. One possibility is to use the random number generator included in the Analysis ToolPak.

Figure 18-6 shows the window produced after you select Random Number Generation in the Analysis ToolPak window.

In this example, the following information is entered into the dialog box. The number of variables equals 1, since the random numbers are being used to compute different possible values for a single variable, the stock price. The number of random numbers being generated is 100. You choose the normal distribution from the Distribution drop-down menu.

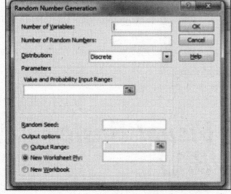

Figure 18-6:
The Random Number Generation window.

Source: Alan Anderson

Once you choose the normal distribution, the window prompts for the appropriate inputs. In this case, the inputs are the mean and the standard deviation of the normal distribution. Because the random numbers are meant to be chosen from the standard normal distribution, the mean is set equal to 0 and the standard deviation is set equal to 1.

The random seed is an optional value and is needed only if this process will be repeated more than once. You use this field to save different sets of random numbers.

The output options allow you to choose from the following options:

- ✔ Have the results written on the same worksheet.
- ✔ Have the results written to a new worksheet in the same Excel workbook.
- ✔ Have the results written to an entirely new Excel workbook.

Figure 18-7 shows the inputs used for this example.

When you've entered all needed parameters, click OK.

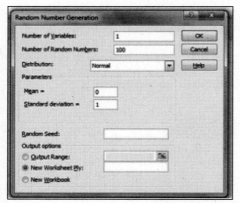

Figure 18-7: Generating 100 standard normal random numbers with the Analysis ToolPak.

Source: Alan Anderson

Programming with Visual Basic for Applications (VBA)

Some applications may require more advanced statistical or data preparation procedures than the ones that are built in to Excel or included with the Analysis ToolPak. Or it may be necessary to perform statistical analyses much more quickly than is possible in Excel. In these cases, you can extend

Excel's capabilities with a built-in programming language known as Visual Basic for Applications (VBA).

You access the VBA language from the Developer tab. When Excel is first installed on a computer, the Developer tab doesn't automatically appear. Follow these steps to add the Developer tab to the other tabs in the menu bar:

1. **Click the File tab.**

2. **Click Options.**

3. **Click Customize Ribbon.**

4. **On the right-hand side of the dialog box, check the Developer check box.**

Once you have installed the Developer tab, choosing it will produce a specialized ribbon that contains the VBA editor.

Figure 18-8 shows the Developer ribbon.

Figure 18-8: The Developer ribbon.

Source: Alan Anderson

To write a new VBA program, the first step is to choose the Visual Basic icon on the Developer ribbon. Figure 18-9 shows a section of the VBA editor.

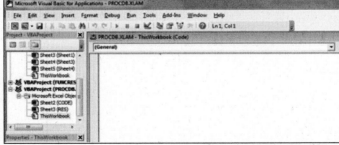

Figure 18-9: The VBA editor.

Source: Alan Anderson

To write a new program, select the Insert tab and choose Module. A blank module will appear in the VBA editor. You type all the code in the module. For example, suppose an investor wants to compute the volatility of a stock based on pricing data from the previous 12 months. The data is shown in Table 18-5.

Table 18-5	Month Stock Price Data
Month	**Prices**
January 2015	100.00
February 2015	101.25
March 2015	103.89
April 2015	104.15
May 2015	102.88
June 2015	103.15
July 2015	109.08
August 2015	107.66
September 2015	105.38
October 2015	104.93
November 2015	106.92
December 2015	107.63

The VBA code is shown in Figure 18-10.

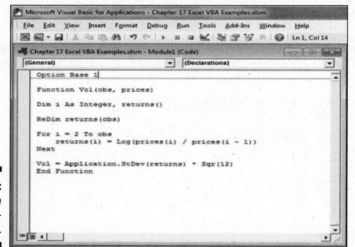

Figure 18-10: VBA code for computing volatility.

The program reads in the 12 monthly prices, computes monthly returns, and then calculates the standard deviation (volatility) of these returns. It also converts the monthly volatility into an annual volatility using the *square root of time rule*.

According to the square root of time rule, monthly volatility can be converted to annual volatility by multiplying the monthly volatility by the square root of 12 (because there are 12 months in a year).

The following is an explanation of the code:

- **Option Base 1:** This command ensures that the index used with any array variables starts with 1. By default, the index starts with 0.

- **Function:** This command gives the name of the function and shows any required inputs. In this case, the required inputs are the number of observations (obs) and the monthly prices.

- **Dim:** This command tells VBA the names and types of the variables and constants that will be used in the program.

- **ReDim:** This command tells VBA the dimensions of any array variables being used. In this case, the returns are stored in an array.

- **For/To/Next:** These commands create a "for" loop. They tell VBA to repeat the commands contained within the loop a specified number of times. In this case, the index i ranges from 2 to obs (2 to 12). Therefore, the loop will be executed 11 times.

- **returns(i) = Log(prices(i) / prices(i – 1)):** This command tells VBA that the continuously compounded monthly return equals the natural logarithm of the current month's price divided by the previous month's price. Because this command appears in the *for loop,* it will be repeated 11 times, once for each return. Note that Log is used in VBA instead of ln, which is used in Excel for the natural logarithm.

- **Vol = Application.StDev(returns) * Sqr(12):** This command computes the value of the volatility that will appear on the spreadsheet. The standard deviation of the monthly returns is multiplied by 12 to produce the annual volatility. The keyword *Application* is used to tell VBA that the function StDev is found in Excel; it is not one of VBA's own commands. Note that Sqr is used in VBA instead of sqrt, which is used in Excel for the square root.

In this example, the final result is a volatility of 0.074768, or 7.4768 percent per year.

All VBA keywords appear in blue in the code. These words have a specific meaning in VBA and cannot be used for any other purposes.

R, Matey!

Whereas VBA is a general, all-purpose programming language, R is designed specifically for statistical analysis. You can download it from the Internet — for free! R becomes far more user-friendly when combined with an interface known as R Studio. Luckily, R Studio is also free on the Internet.

To download the R language, go to www.r-project.org; both Windows and Mac versions are available (and Linux is available as well). You can download R Studio from www.rstudio.com. Be sure to install R before R Studio.

The R language evolved from an earlier language known as S, which was developed at Bell Laboratories in the 1970s. With R, you can generate results with the built-in functions that are part of the R language, or you can write you own code. One of the main advantages of R is that it can easily be extended with add-in packages. A large network of R users, including statisticians, economists, computer programmers, and mathematicians regularly contribute powerful new packages, all of which can be downloaded online at no cost.

Figure 18-11 shows the layout of R Studio. R Studio is designed to make the R language user-friendly and enables users to keep track of several different types of information at once.

Figure 18-11:
R Studio.

Source: Alan Anderson

In the upper left-hand corner of R Studio is an editor you can use to enter code without executing it. Once the code is complete, you execute it by highlighting the relevant lines and clicking the Run button. The editor window can also be used for several other purposes. For example, you can use it to retrieve information about R functions, create comments for documenting code, and so on.

The R Console is in the lower left-hand corner of R Studio. This is where the results of executed code appear. You can also enter R code directly into the R Console for immediate execution. One convenient feature of the console is that you can retrieve past commands by using the up-arrow key; this saves time entering commands and reduces the likelihood of errors.

In the upper right-hand corner of R Studio is a window containing the environment and history tabs. The Environment window shows the variables that are currently defined in R's memory. It also shows the type of each variable (numerical, character, and so on) and its dimensions (the number of rows and columns). Clicking the name of a variable produces a chart in the editor window that contains the current value(s) of the variable.

The History window shows the R code that has been used during the current session. This code can be executed by selecting it and then clicking the To Console button. Also, you can send it to the editor by clicking the To Source button.

The bottom right-hand corner of R Studio contains several useful tabs:

- Files
- Plots
- Packages
- Help

The Files window shows the contents of the directory that contains the R software.

The Packages window shows the add-in packages that have been downloaded and are available to R. These are divided into two groups: Those that are part of the R language are listed under System Library, and the rest are listed under User Library. Packages in this window must be activated by clicking the check box next to the package's name or using the appropriate command in an R program.

The Help window contains a massive amount of information about the R language. Also, you can get help for any command by typing **??** and the name of the command in the R Console.

For example, R contains a dataset that consists of the number of times (in minutes) between eruptions of the geyser Old Faithful. Figure 18-12 shows the results of applying several statistical functions to the data.

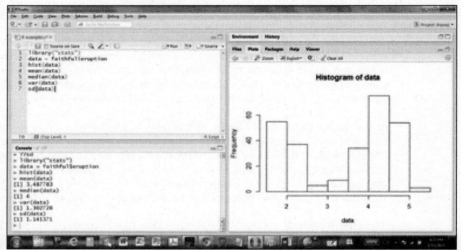

Source: Alan Anderson

Figure 18-12:
Basic
statistical
commands
in R.

Here is the code:

```
library("stats")
data = faithful$eruption
hist(data)
mean(data)
median(data)
var(data)
sd(data)
```

The code begins with the command `library("stats")`. The `library` command is used to activate an add-in package so it can be used by the program. The stats package contains a large number of useful statistical functions.

The command `data = faithful$eruption` assigns the contents of the Old Faithful dataset to the variable called `data`. This is done to increase the readability of the code.

The hist command creates a histogram of the data. Figure 18-13 shows the histogram; it's used to describe the properties of a dataset by showing a bar for each individual value or range of values on the horizontal axis. The corresponding frequencies are represented by the heights of the bars. The histogram for the Old Faithful data shows that waiting times between 4 and 4.5 minutes are the most common; waiting times between 2.5 and 3 minutes are the least common. The data also does not appear to follow any particular probability distribution such as the binomial or the normal.

The remaining commands compute statistical measures for the data: mean, median, sample variance, and sample standard deviation. The results appear in the R Console.

For the Old Faithful data, the mean waiting time is 3.487783 minutes, and the median time is 4 minutes. This indicates that half of the waiting times are less than 4 minutes, and the rest are greater than 4 minutes. The variance of the waiting times is 1.302728, and the standard deviation is 1.141371 minutes. (The variance is measured in squared minutes.)

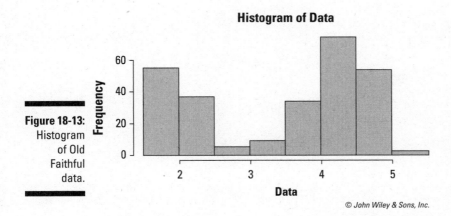

Figure 18-13: Histogram of Old Faithful data.

© John Wiley & Sons, Inc.

So as you can see, even if you don't have an industrial-strength statistical package at your disposal, there is a great deal you can do with free computer-based packages. True big-data applications in industry require more firepower, but you can build a good deal of intuition and understanding of statistical procedures using the tools described in this chapter.

Chapter 19

Seeking Free Sources of Financial Data

. . . .

In This Chapter

▶ Checking out what types of financial data are available on the Internet

▶ Downloading financial data from various Internet sources

. . . .

So far, this book has focused on financial applications. Most of the examples have been in the vein of modeling returns on various types of assets. As pointed out throughout the chapters in Part I of this book, big-data applications arise in a vastly diverse collection of circumstances. Because of this diversity, it is almost impossible to give an exhaustive list of data sources. The data you use will depend completely on your particular situation.

So, this chapter aims to give you some resources that provide financial data. Our thought is that this will help you to get your feet wet as you try to apply some of the techniques in this book to actual real-world data.

You can find a large number of sources of financial data on the Internet. In many cases, you can download the data into a spreadsheet — free of charge. The available data includes stock prices, option prices, exchange rates, interest rates, and so forth. There are also key financial measures for individual corporations, such as revenues, profits, and the price/earnings (P/E) ratio. Many websites offer a wealth of financial news and analysis as well.

This chapter focuses on four of the best sources of free economic and financial data:

- ✔ Yahoo! Finance
- ✔ Federal Reserve Economic Data (FRED)
- ✔ Board of Governors of the Federal Reserve System
- ✔ U. S. Department of the Treasury

Yahoo! Finance

The Yahoo! Finance website at `http://finance.yahoo.com` is one of the best online sources of stock prices and corporate financial data. In addition, the website provides useful information about the following topics:

✔ Option trading strategies

✔ Personal finance

✔ Economic news

✔ Company news

The ticker symbol

Finding data about a stock or a corporation on Yahoo! Finance requires that you use stock's *ticker symbol,* a series of letters and/or numbers that uniquely identifies a stock. For example, AAPL represents Apple stock, which trades on the NASDAQ (National Association of Securities Dealer Automated Quotation) system. Table 19-1 shows a sample of ticker symbols for other widely traded stocks (note that NYSE stands for the New York Stock Exchange).

Table 19-1	Stock Ticker Symbols	
Stock	*Symbol*	*Exchange*
AT&T	T	NYSE
Bank of America	BAC	NYSE
Citigroup	C	NYSE
ExxonMobil	XOM	NYSE
Ford Motor Co.	F	NYSE
General Electric	GE	NYSE
Microsoft	MSFT	NASDAQ
Oracle	ORCL	NYSE
Twitter	TWTR	NYSE
Walmart	WMT	NYSE

Historical origins of ticker symbols

Ticker symbols and corresponding stock prices were originally printed by a machine known as a *stock ticker.* That's an invention that was used for a century, from the 1870s until 1970. Thomas Edison invented a good one — the Universal Stock Ticker — in 1869; it used alphanumeric characters instead of Morse code.

Data was transmitted from stock exchanges through telegraph wires to the stock ticker machines, so obtaining this data was very expensive. The machines printed transactions on a thin strip of paper known as *ticker tape.* And of course, ticker tape was traditionally tossed out of the windows of office buildings during parades, which became known as *ticker tape parades.* Because ticker tape was very narrow, it was important for ticker symbols to be as compact as possible.

You can use a search engine such as Google to find the symbol for virtually any stock traded in the United States. For example, to find the ticker symbol for ExxonMobil, just type "ticker symbol ExxonMobil" into Google; you immediately get a results display that includes the symbol, recent selling price, and recent volume. You can also click on links that take you to related information about ExxonMobil.

Alternatively, it is possible to search for ticker symbols on `finance.yahoo.com`. You type the first few letters of a corporation's name into the dialog box at the top of the home page, and a list of suggestions appears. Clicking on the correct ticker symbol brings up the appropriate page for that stock.

Downloading historical stock prices

Once you find the page for a given stock, several types of data are available for download. The first column on the far left of the page offers some choices for data about the stock, including historical prices, charts, company news, and financial statements.

To download historical prices, click the Historical Prices link to bring up the Historical Prices page.

In the Set Date Range section of the page, enter the start date and end date for the desired data. There are several frequencies for the data: daily, weekly and monthly. It's also possible to choose dividends instead of prices.

Clicking the Get Prices button results in a list of fields for each date in the chosen range. These fields are: Open Price, High Price, Low Price, Close Price, Volume, and Adjusted Close.

- ✔ **Open Price** is the price at which the stock began trading.
- ✔ **High Price** is the highest price that the stock reached during the trading day.
- ✔ **Low Price** is the lowest price that the stock reached during the trading day.
- ✔ **Close Price** is the last price at which the stock traded.
- ✔ **Volume** refers to the number of shares that traded during the day.
- ✔ **Adjusted Close** is the closing price after adjusting for dividends or any other cash flows that took place prior to the next trading day.

After clicking the Get Prices button, the selected data comes up.

At the bottom of the page you'll find the Download to Spreadsheet link. Clicking this link produces a spreadsheet containing the data. You can then save that as a .CSV (comma separated variables) file or as an .XLSX (Excel) file.

The .CSV format consists of data in which the values are separated by commas. The .CSV format doesn't offer as many features as the .XLSX format, but is compatible with many different types of software in addition to Excel. You can open or import any .CSV file into Excel and save it in the .XLSX format.

Finding stock option prices

On a stock page's main menu, the first main category is Quotes. Under this category is a link called Options. Clicking Options produces a page showing the prices of the calls and puts that are available for the stock.

Owning a *call* gives you the right to *buy* a stock at a specified price, known as the *strike price* (or *exercise price*). The call option matures at a specified time in the future. If the call isn't *exercised* by then (that is, you don't purchase the stock), the option *expires* and becomes worthless. A *put* gives you the right to *sell* a stock at a specified strike price, and if it's not exercised also becomes worthless after the maturity date passes. The two basic option styles are *European,* which may only be exercised on the maturity date, and *American,* which may be exercised at any time until the option matures.

By default, the options page is split up so that calls are at the top and puts are at the bottom. You can change this by clicking on the Straddle button on the top of the page. Doing so causes call prices to be shown on the left of the screen and put prices to be shown on the right.

Calls and puts on a stock are typically available with several different strike prices and maturity dates. For each option, the website shows Ticker Symbol, Last Price, Change, Bid Price, Ask Price, Volume, and Open Interest:

- ✔ **Last Price** is the last price at which the option traded.

- ✔ **Change** represents the change in the option's last price since the previous trading day.

- ✔ **Bid** is the "bid price" of the option. If an investor chooses to sell an option to a dealer, he or she will receive the bid price.

- ✔ **Ask** is the "ask price" of the option. If an investor chooses to buy an option from a dealer, he or she will pay the ask price. The difference between the bid and ask prices is known as the "bid-ask" spread, and that's the source of a dealer's profits.

- ✔ **Volume** represents the number of options that were traded during the day.

- ✔ **Open Interest** represents the number of outstanding options that have not yet reached their maturity date.

Looking at the option prices, you'll notice that the calls with the lowest strike prices are the most expensive, and for puts, the pattern is reversed. For both calls and puts, a longer maturity option is more expensive than a shorter maturity option.

Analyzing options strategies

Yahoo! Finance gives investors the ability to analyze several different option trading *strategies*, including the following:

- ✔ Butterfly spreads
- ✔ Spread trades
- ✔ Straddles

Options may be combined in many interesting ways in order to *hedge* existing positions or *speculate* on the future behavior of an asset price, interest rate, or index. Investors hedge to reduce or eliminate the risk of a position in order to (hopefully) make a profit.

You can hedge the risk that the price of an asset or commodity will rise in the future by buying a call option. You can hedge the risk that the price will fall by buying a put option. The call option guarantees your lower buying price, even if the market price rises. The put option guarantees your higher selling price, even if the market price drops.

As an example of a trading strategy, if a trader believes that the price of a stock will remain stable over the next three months, he or she can set up a *butterfly spread* in order to profit from this view. This strategy will be profitable if the price of the stock doesn't fluctuate a great deal, with the added benefit that the trader's losses cannot exceed the cost of setting up the position.

The butterfly spread strategy essentially involves buying options with strike prices above and below the current market price of the stock and selling options at the current market price. There are a variety of possible scenarios from there, depending on what the stock price does. But in the worst case, you can use the options you bought to cover your exposure on the ones you sold. In the best case, if the stock price doesn't change at all, the options you sold expire worthless and you pocket the selling price. In this case, one of the options you bought will also be in the money.

Options strategies provide stock investors with a great deal of flexibility — flexibility that isn't usually possible with other assets.

You can find several different options strategies by going to the Options Center page. To get there, follow these steps (bearing in mind that websites change their designs all the time):

1. **Enter a stock ticker symbol on the home page for Yahoo! Finance and then click the Options link on the main menu to bring up the Options page.**

2. **Click the Market Data link (near the top of the page, directly below the dialog box) to bring up the Market Data page.**

   Hovering over the Market Data link on the left-hand side of the page produces a pop-up menu. From there, you can choose the Options link to bring up the Options Center page.

   Directly beneath the Options Center title is a link titled Options Analysis Tool. The options strategies are listed under the section Options Strategies on the left-hand side of the page. Each strategy is analyzed in great detail and provides links to other relevant information about options.

Finding key statistics for a corporation

To properly analyze a stock, you need an understanding of the issuing corporation's *financial statements*. The main financial statements are the income statement, the balance sheet, and the statement of cash flows.

The *income statement* shows the revenues and expenses incurred by a corporation during a given year, and as a result its profits (or losses) during the same period. The *balance sheet* shows the position (assets, liabilities, and equity) of a corporation at a given point in time. The *cash flow statement* is a detailed look at the corporation's receipts of cash and its payments of cash during a given time period.

To find the financial statements for a corporation, the first step is to find its stock page using the stock's ticker number. On the stock's page, look for the category Company on the main menu at the left-hand side of the screen. Then click the Key Statistics link to bring up a page that summarizes a large number of statistics from the corporation's financial statements. These include market capitalization, price-earnings ratio, revenue, gross profit, and so forth. The same page also has details about the performance of the corporation's stock over the past 52 weeks, including the range of prices (the maximum and the minimum stock selling price), details about dividends, and so forth.

You can find more detail under the category Financials on the left-hand side of the page. Here are the appropriate links:

- Income Statement
- Balance Sheet
- Cash Flow

These three statements contain detailed information about the performance of the corporation. The SEC (United States Securities and Exchange Commission) requires corporations to file these documents every year.

Other information on Yahoo! Finance

Yahoo! Finance also provides data about many other financial assets, such as mutual funds, exchange-traded funds (ETFs), bonds, commodities, and currencies. You can find all of these by clicking the Market Data link on the left-hand side of the home page.

The mutual funds page shows a list of the best performing mutual funds. (Recall that a *mutual fund* is a collection of stocks that provides the benefits of diversification without requiring an investor to choose individual stocks.)

The *exchange-traded fund* (ETF) page shows the top performing ETFs. ETFs are similar to mutual funds except that you can trade them like a single stock, whereas mutual funds are typically more restrictive. And because the ETFs are based on a fixed set of stocks (like the S&P 500, for example), you're typically not paying for the services of a portfolio manager.

The bonds page summarizes the yields of currently traded U.S. Treasury securities. This page also has links to other bonds, such as municipal bonds (issued by city, state, and local governments) and corporate bonds.

The relationship between the yields and maturities of Treasury securities is known as the *term structure of interest rates*. The *yield curve* is a graph of this relationship.

The commodities page mainly shows *futures* prices for crude oil, heating oil, and natural gas.

A *futures contract* is a derivative security that allows an investor to lock in the price at which he or she may buy or sell an asset in the future. Futures contracts trade on organized exchanges and are often used for hedging purposes. The price of an asset that will be delivered through a futures contract is known as the *futures price*.

The currencies page shows the exchange rates between the major currencies, such as the dollar, euro, British pound, Japanese yen, Canadian dollar, Swiss franc, and so forth.

Federal Reserve Economic Data (FRED)

The Federal Reserve Bank of St. Louis provides an economic database free of charge to the public at its website `http://research.stlouisfed.org/fred2/`. The database is known as FRED (Federal Reserve Economic Data). Note that this website requires you to sign up for a free account in order to access the data.

The Federal Reserve System is the central bank of the United States. There are 12 district Federal Reserve Banks, and FRED is maintained by the St. Louis Federal Reserve Bank. The primary purpose of the Federal Reserve is to manage the money supply in the U.S.

Unlike Yahoo! Finance, FRED provides federal financial data and macroeconomic data, including the following:

- Gross Domestic Product (GDP) is the value of all goods and services produced in the United States during a given year.

✔ The federal funds rate (also known as the *fed funds rate*) is the interest rate charged by banks for overnight loans from other banks.

✔ The Consumer Price Index (CPI) is a measure of the cost of goods and services bought by consumers during the year.

Finding macroeconomic data on FRED

To locate macroeconomic data, first click the Category link at the top of the home page. This produces a page showing the different categories of data. For example, to locate the consumer price index (CPI), click Consumer Prices (CPI and PCE) under the Prices category. Several types of price indexes are available. To download the CPI, choose Consumer Price Index for All Urban Consumers: All Items. You may choose seasonally adjusted data or non-seasonally adjusted data.

Once you've selected one or more data series, click the Add to Data List button. This brings you to the Add to My Data List page. On this page, you're offered many types of data. For example, *levels* refers to the actual CPI; *change* and *percentage change* show how the CPI has changed since the previous trading day.

After clicking the type of data you want, click the Add Series to My Data List button. This brings you to the Data List page, where clicking the Download Data button brings you to the Download Data page. From there, you're offered the choice of downloading the data as an Excel file or as a *tab-delimited file.* You're also offered the chance to specify a range of dates for the data.

A tab-delimited file is similar to a comma-separated file, except the data fields are separated by tabs instead of commas. Also, because the data files you're downloading are potentially very large, they're stored in a zip format. A zip file is one that's been compressed in order to save space. To read a zip file, it must be unzipped. Double-clicking a zipped file will usually unzip it on Windows and Mac machines.

Finding financial data on FRED

FRED also contains the following types of financial data:

✔ Exchange rates

✔ Interest rates

✔ Bond market indices

✔ Stock market indices

To locate financial data, the first step is to click the Categories link at the top of the home page. Financial data is under the Money, Banking & Finance category.

For example, to find exchange rate data, click on the Exchange Rates link under the Money, Banking & Finance category. There are several choices within this category, including daily rates, monthly rates, and annual rates. Choosing one of these categories provides several more choices, such as the exchange rate between the U.S. dollar and the euro, the Japanese yen, the Chinese yuan, the Canadian dollar, and several other foreign currencies. Data for each currency is referred to as a *data series*. You may download as many as you like by clicking the check box next to each item.

The remaining steps are the same as for downloading macroeconomic data. Click the Add to Data List button to bring you to the Add to My Data List page. On this page, you're offered many types of data for a specified exchange rate. For example, *levels* refers to the actual exchange rate, and *change and percentage change* show how an exchange rate has changed since the previous trading day.

After clicking the type of data you want, click the Add Series to My Data List button. This brings you to the Data List page, where clicking the Download Data button brings you to the Download Data page. From here, you're offered the choice of downloading the data you've selected as an Excel file or a tab-delimited file.

Board of Governors of the Federal Reserve System

The Board of Governors of the Federal Reserve System website at http://federalreserve.gov provides a great deal of information about the conduct of monetary policy in the United States, along with economic and financial data. The Board of Governors oversees the Federal Reserve System. The board is responsible for regulating the banking system, is involved in the conduct of monetary policy, and produces economic data.

You can find economic and financial data by clicking the Economic Research & Data tab. From the Economic Research and Data page, click the Data Releases link. On the Data Releases page, several categories of data appear, including exchange rates, interest rates, bank finance data, and household finance data.

For example, under the Interest Rates category, there's a link to Selected Interest Rates — the available frequencies are daily and weekly. Here are some of the rates that are available:

- ✔ Commercial paper is a short-term promissory note issued by a corporation to raise funds.

- ✔ U.S. Treasury securities are bonds issued by the United States Department of the Treasury in order to raise funds for the government.

- ✔ Eurodollar deposits are dollar deposits held at foreign banks or foreign branches of U.S. banks.

- ✔ Interest rate swaps are derivative securities in which cash flows are exchanged, based on the difference between two interest rates — a *fixed* rate and a *floating* rate (that is, one that is reset periodically).

The Board of Governors website also contains data for the United States money supply and industrial output, and data about the assets and liabilities of United States banks.

U.S. Department of the Treasury

The United States Department of the Treasury website at www.treasury.gov provides a great deal of information about Treasury securities, along with their historical yields.

The Treasury Department is responsible for raising revenues for the U.S. government. It is the home of the Internal Revenue Service and also issues coins, which are produced by the U.S. Mint. (Currently, all coins meant for general circulation are produced by the Denver mint and the Philadelphia mint.) The Treasury Department also oversees the banking system and enforces tax laws.

In the Data Center box at the bottom of the home page, click the Data Center link to bring up the Resource Center page. Under the Interest Rates category, data is available for Daily Treasury Yield Curve Rates, Daily Treasury Real Yield Curve Rates, Daily Treasury Bill Rates, and Daily Long-Term Rates.

Treasury securities can be classified as follows: Treasury bills have a maturity of up to 1 year, Treasury notes have a maturity between 2 and 10 years, and Treasury bonds have a maturity between 20 and 30 years.

Real rates of interest have been adjusted for inflation; nominal rates of interest (the ones observed every day) have *not* been adjusted for inflation.

If you're interested in data about daily Treasury yield curve rates, click the Daily Treasury Yield Curve Rates link. This brings up a page showing current yields for Treasury securities with maturities ranging from 1 month to 30 years. To get historical rates, choose the appropriate year in the dialog box labeled Select Time Period and click Go. This shows the selected data on the web page. To download this data into Excel, you can copy and paste the data on the web page.

The website also provides a great deal of information about economic policy, tax policy, small business programs, and international institutions such as the International Monetary Fund (IMF).

Other Useful Financial Websites

There are many other websites where financial news, analysis, and data may be found. These include:

- NASDAQ (www.nasdaq.com)
- Seeking Alpha (www.seekingalpha.com)
- MarketWatch (www.marketwatch.com)
- Morningstar (www.morningstar.com)
- Bloomberg (www.bloomberg.com)

The NASDAQ website provides stock prices, option prices, and market indexes, as well as news about stocks. It also provides detailed information for investors, such as investment basics and details about investing in more sophisticated securities such as ETFs, currencies, bonds, commodities, and options. It also contains useful information about personal finance.

Seeking Alpha is intended for sophisticated investors. It provides a large number of articles drawn from the Internet about different investment topics, news articles, and stock market information. It also covers other investments such as currencies, ETFs, and so forth.

MarketWatch, Morningstar, and Bloomberg also provide investors with a substantial amount of financial news and data about financial assets such as stocks, bonds, currencies, and so on.

Part V
The Part of Tens

the

part of

tens

There's a free extra Part of Tens chapter at www.dummies.com/extras/statisticsforbigdata.

In this part . . .

- ✔ Following best practices in data preparation
- ✔ Answering ten common questions about EDA

Chapter 20

Ten (or So) Best Practices in Data Preparation

In This Chapter
▶ Understanding the key steps in data validation
▶ Preparing data for analysis

*T*he main goal of this book is to get you familiar with the statistical methods that allow you to build useful statistical models. But as you've probably noticed, we have spent a great deal of time, particularly in Part II, talking about getting data ready for analysis. Statistical software packages are extremely powerful these days, but they cannot overcome poor quality data. This chapter provides a checklist of things you need to do before you go off building statistical models.

Check Data Formats

Your analysis always starts with a raw data file. Raw data files come in many different shapes and sizes. Mainframe data is different than PC data, spreadsheet data is formatted differently than web data, and so forth. And in the age of big data, you will surely be faced with data from a variety of sources. Your first step in analyzing your data is making sure you can read the files you're given. Chapter 7 gives some tips about how to do this.

Chapter 6 talks about the formats of the individual data fields, or variables, in your data file. You need to actually look at what each field contains. For example, it's not wise to trust that just because a field is listed as a character field, it actually contains character data.

Be particularly careful with date formats. Different software packages treat dates differently. Often in statistical analysis, you calculate elapsed time between two dates. This calculation may require some massaging of your date data to produce a meaningful result. Chapter 6 gives you detailed information about how to deal with dates.

Verify Data Types

All data falls into one of four categories that affect what sort of statistics you can appropriately apply to it:

- ✔ Nominal data is essentially just a name or an identifier.
- ✔ Ordinal data puts records into order from lowest to highest.
- ✔ Interval data represents values where the differences between them are comparable.
- ✔ Ratio data is like interval data except that it also allows for a value of 0.

Chapter 5 discusses these categories in detail and explains what statistical procedures you can apply to which categories of data. It's important to understand which categories your data falls into before you feed it into the statistical software. Otherwise, you risk ending up with perfectly reasonable-looking gibberish.

Graph Your Data

Getting a sense of how your data is distributed is important. You can run statistical procedures until you're blue in the face, but none of them will give you as much insight into what your data looks like as a simple graph.

Chapters 11 and 12 talk extensively about data visualization. You can use graphical techniques for many of the data preparation steps discussed in this chapter. In particular, the next four sections describe processes that could benefit from graphical techniques.

Verify Data Accuracy

Once you're comfortable that the data is formatted the way you want it, you still need to make sure it's accurate and that it makes sense. This step requires that you have some knowledge of the subject area you are working in.

There isn't really a cut-and-dried approach to verifying data accuracy, as discussed in Chapter 6. The basic idea is to formulate some properties that you think the data should exhibit and test the data to see if those properties hold. Are stock prices always positive? Do all the product codes match the list of valid ones? Essentially, you're trying to figure out whether the data really is what you've been told it is.

Identify Outliers

Outliers are data points that are out of whack with the rest of the data. They are either very large or very small values compared with the rest of the dataset.

Outliers are problematic because they can seriously compromise statistics and statistical procedures. Chapter 10 (about outliers) presents an example that shows how a single outlier can have a huge impact on the value of the mean. Because the mean is supposed to represent the center of the data, in a sense, this one outlier renders the mean useless.

When faced with outliers, the most common strategy is to delete them. In some cases, though, you may want to take them into account. In these cases, it's usually desirable to do your analysis twice — once with outliers included and once with the outliers excluded. This allows you to evaluate which method gives more useful results.

Deal with Missing Values

Missing values are one of the most common (and annoying) data problems you will encounter. Your first impulse might be to drop records with missing values from your analysis. The problem with this is that missing values are frequently not just random little data glitches.

Chapter 9, which deals with missing values, distinguishes between several different scenarios involving missing values. The primary distinction is between data that's missing randomly and data that's missing in some systematic way. In the latter case, deleting these records may significantly change the nature of the data and therefore affect the analysis.

Chapter 9 goes on to discuss some ways of supplying missing data. It describes several techniques for making educated guesses about what the missing data would actually be. You do this by looking for patterns in the records where data isn't missing. In this way, you can preserve the whole dataset and analyze it usefully.

Check Your Assumptions about How the Data Is Distributed

Many statistical procedures depend on the assumption that the data is distributed in a certain way. If that assumption fails to be the case, the accuracy of your predictions suffers.

The most common assumption for the modeling techniques discussed in this book is that the data is normally distributed. In Chapter 8, which talks about various ways of checking this assumption, points out that it's possible to quite rigorously verify the assumption of a normal distribution.

Or not. In cases where the data isn't distributed as you need it to be, all is not necessarily lost. There are a variety of ways of transforming data to get the distribution into the shape you need it, some of which are explored in Chapters 15 and 16.

One of the best ways to verify the accuracy of a statistical model is to actually test it against the data once it's built. One way to do that is to randomly split your dataset into two files. You might call these files Analysis and Test, respectively.

You need to split the data randomly to be effective. You cannot simply split the dataset into the top half and the bottom half, for example. Almost all data files are sorted somehow — by date if nothing else. This introduces systematic patterns that will give different portions of the file different statistical properties. When you split the file randomly, you give each record an equal chance of being in either file. Figuratively, you're flipping a coin for each record to decide which file it goes into. Randomness gives both files the same statistical properties as the original data.

Once you have split the dataset, set aside the Test file. Then proceed to build your predictive model using the Analysis file. Once the model is built, apply it to the Test file and see how it does.

Testing models in this way helps safeguard against a phenomenon known as *over-fitting*. Essentially, it's possible for statistical procedures to memorize the data file rather than discover meaningful relationships among the variables. If over-fitting occurs, the model will test quite poorly against the Test file.

Back Up and Document Everything You Do

Because statistical software is getting to be so simple to use, it's a piece of cake to start generating reports and graphs, not to mention data files. You can run procedures literally at the touch of a button. You can generate several dozen graphs based on different data transformations in a matter of a few minutes. That makes it pretty easy to lose track of what you have done, and why.

It's important to make sure you keep a written record of what you're up to. Graphs should be labeled with the name (and version) of the data that was used to create them. Statistical procedures that you build need to be saved and documented.

It's also important to back up your data files. In the course of your analysis, you will likely create several versions of your data that reflect various corrections and transformation of variables. You should save the procedures that created these versions. They should also be documented in a way that describes what transformations you have made and why.

Documentation isn't anyone's favorite task, but we speak from experience when we strongly encourage you not to rely on your memory when it comes to your analysis projects.

By working through the steps just described, you maximize the reliability of your statistical models. In many cases, the prep work is actually more time-consuming than the actual model building. But it's necessary. And you'll thank yourself in the end for working through it methodically.

Chapter 21

Ten (or So) Questions Answered by Exploratory Data Analysis (EDA)

In This Chapter

▶ Understanding the most important questions answered by Exploratory Data Analysis (EDA)

▶ Seeing how to use EDA to determine if a dataset conforms to your assumptions

*T*his chapter covers ten key questions about a dataset that can be answered by using exploratory data analysis (EDA). These questions focus on the statistical properties of the data, along with the distribution followed by the data and the nature of the relationships among the variables in the data.

What Are the Key Properties of a Dataset?

Prior to performing any type of statistical analysis, understanding the nature of the data being analyzed is essential. You can use EDA to identify the properties of a dataset to determine the most appropriate statistical methods to apply to the data. You can investigate several types of properties with EDA techniques, including the following:

✔ The center of the data

✔ The spread among the members of the data

✔ The skewness of the data

✔ The probability distribution the data follows

> ✔ The correlation among the elements in the dataset
>
> ✔ Whether or not the parameters of the data are constant over time
>
> ✔ The presence of outliers in the data

Chapter 5 introduces most of these notions. Chapter 16 talks about constancy over time in its discussion of time series. Outliers are the subject of Chapter 10.

Another key question EDA answers is "Does the data conform to our assumptions?"

Identifying the properties of a dataset is very important, because many statistical procedures are sensitive to the assumptions you make about the data.

What's the Center of the Data?

You identify the center of a dataset with several different *summary measures*. These include the big three:

> ✔ Mean
>
> ✔ Median
>
> ✔ Mode

You calculate the *mean* of a dataset by adding up the values of all the elements and dividing by the total number of elements. For example, suppose a small dataset consists of the number of days required to receive a package by the residents of an apartment complex:

1, 2, 2, 4, 7, 9, 10

The mean of this dataset would be the following:

$$\frac{1+2+2+4+7+9+10}{7} = \frac{35}{7} = 5$$

The average length of time for the residents to receive a package is 5 days.

The *median* of a dataset is a value that divides the data in half. The first half contains the smallest elements and the second half consists of the largest elements. In the previous example, because the data consist of seven observations, the fourth smallest value would be the median:

1, 2, 2, *4*, 7, 9, 10

The median is 4, because half of the observations are less than 4, and half are greater than 4.

The *mode* of a dataset is simply the most frequently occurring value. With the package delivery example, the mode is 2.

For a real-world example, Figure 21-1 shows a histogram for daily returns to ExxonMobil stock in 2013.

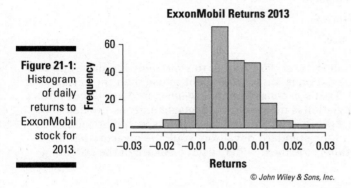

Figure 21-1: Histogram of daily returns to ExxonMobil stock for 2013.

© John Wiley & Sons, Inc.

Each bar represents a *range* of values; the width of each interval is 0.005. The heights of the bars indicate how many returns fell within each interval. The histogram makes it easy to see which ranges of values occurred the most frequently and which occurred the most infrequently.

The histogram shows that most of the returns are close to the mean, which is 0.000632 (0.0632 percent). The median is –0.000118, and the mode could be considered to be the range of values between –0.005 and 0.

How Much Spread Is There in the Data?

You identify the spread of a dataset from the center with several different summary measures:

- ✔ Variance
- ✔ Standard deviation
- ✔ Quartiles
- ✔ Interquartile range (IQR)

Variance is the average squared deviation between the elements of the dataset and the mean. For a sample of data, the variance is computed like this:

$$s^2 = \frac{\sum_{i=1}^{n}(x_i - \bar{x})^2}{n-1}$$

where

x_i is the value of a single element in the sample.

$\bar{x}$ is the sample mean.

n is the sample size.

The standard deviation is the square root of the variance. For most applications, the standard deviation is more convenient to use than the variance as a measure of spread. That's because variance is measured in *squared* units, whereas standard deviation is measured in the same units as the data. For example, the variance of a dataset consisting of prices would be measured in dollars *squared,* and the standard deviation would be measured in dollars. Standard deviation is the most widely used measure of the spread in a dataset.

Quartiles divide a dataset into four equal parts. The first quartile (Q_1) divides the data into the lowest 25 percent of the observations and the highest 75 percent (25 percent of the observations are *less than* Q_1, and 75 percent are *greater than* Q_1). The second quartile (Q_2) divides the data into the lowest 50 percent of the observations and the highest 50 percent. The third quartile (Q_3) divides the data into the lowest 75 percent of the observations and the highest 25 percent. The interquartile range (IQR) equals the difference between the third and first quartiles:

$$IQR = Q_3 - Q_1$$

The IQR represents the middle 50 percent of the data.

The quartiles of a dataset are best illustrated with a *box plot.* Figure 21-2 shows a box plot of the daily returns to ExxonMobil in 2013.

The box plot shows several key statistics for the ExxonMobil returns:

Minimum = -0.0285

$Q_1 = -0.0046$

$Q_2 = -0.0001$

$Q_3 = 0.0064$

Maximum = 0.0284

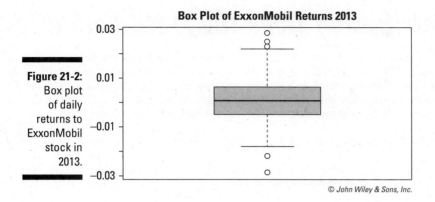

Figure 21-2: Box plot of daily returns to ExxonMobil stock in 2013.

The minimum return is shown on a graph as a single point at the bottom of the plot (a box plot shows *outliers* as individual points). Q_1 is shown as the bottom of the box, Q_2 is the solid black line in the middle of the box, and Q_3 is the top of the box. The maximum return is shown as a single point at the top of the plot.

Is the Data Skewed?

The histogram in Figure 21-1 shows that most of the returns are close to the mean, which is 0.000632 (0.0632 percent). The median is −0.0001179. The relationship between the mean and the median can be used to determine if a distribution is skewed, as follows:

Mean > median = Positively skewed

Mean = median = Symmetrical

Mean < median = Negatively skewed

In this case, the distribution of returns to ExxonMobil stock is positively skewed. This means that large positive returns somewhat outweigh large negative returns in this dataset.

You can also use a histogram to determine if a dataset is skewed. For positively skewed data, the right tail tends to be longer than the left tail. The reverse is true for negative skewed data. In Figure 21-1, the right tail of the distribution appears to be slightly longer than the left tail, which provides more evidence that the data is positively skewed.

What Distribution Does the Data Follow?

One technique you can use to identify the distribution a dataset follows is the QQ-plot (QQ stands for *quantile-quantile,* which we discuss at length in Chapter 12). You can use the QQ-plot to compare a dataset to a large number of different probability distributions. Often, data is compared to the normal distribution because many statistical tests assume normally distributed data.

Figure 21-3 shows a normal QQ-plot for ExxonMobil's daily returns in 2013.

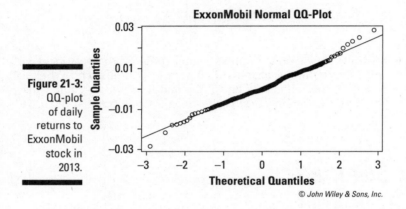

Figure 21-3:
QQ-plot
of daily
returns to
ExxonMobil
stock in
2013.

© John Wiley & Sons, Inc.

The QQ-plot shows the quantiles of the normal distribution on the horizontal axis and the quantiles of the dataset on the vertical axis. If the dataset exactly matches the normal distribution, the points on the graph exactly match the upward-sloping line.

In this case, the returns to ExxonMobil stock closely follow the normal distribution, except for small discrepancies in the left and right tails of the distribution. (The upper right-hand corner of the QQ-plot represents the right tail of the distribution followed by ExxonMobil returns and the bottom left-hand corner represents the left tail.) The QQ-plot shows that the distribution of ExxonMobil returns has slightly fatter left and right tails than the normal distribution.

A more formal statistical test would be required to prove whether or not the ExxonMobil data is normally distributed, but the QQ-plot shows that the data is likely to be normal.

Are the Elements in the Dataset Uncorrelated?

For a dataset that consists of observations taken at different points in time (that is, *time series data*), it's important to determine whether or not the observations are correlated with each other. This is because many techniques for modeling time series data are based on the assumption that the data is uncorrelated with each other (independent).

One graphical technique you can use to see whether the data is uncorrelated with each other is the *autocorrelation function*. The autocorrelation function shows the correlation between observations in a time series with different lags. For example, the correlation between observations with lag 1 refers to the correlation between each individual observation and its previous value.

Figure 21-4 shows the autocorrelation function for ExxonMobil's daily returns in 2013.

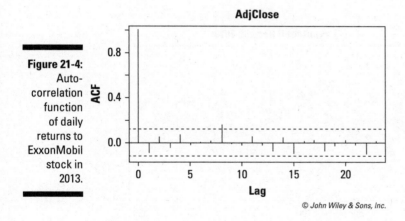

AdjClose

Figure 21-4:
Auto-
correlation
function
of daily
returns to
ExxonMobil
stock in
2013.

© John Wiley & Sons, Inc.

Each "spike" in the autocorrelation function represents the correlation between observations with a given lag.

The autocorrelation with lag 0 always equals 1, because this represents the correlations of the observations with themselves.

On the graph, the dashed lines represent the lower and upper limits of a *confidence interval*. If a spike rises above the upper limit of the confidence interval or falls below the lower limit of the confidence interval, that shows that the correlation for that lag isn't 0. This is evidence against the independence of the elements in a dataset.

In this case, there is only one statistically significant spike (at lag 8). This spike shows that the ExxonMobil returns may be independent. A more formal statistical test would show whether that is true or not.

Does the Center of the Dataset Change Over Time?

For time series data (see Chapter 16), it's important to know whether the observations continue to have the same mean over time. Many statistical tests and forecasting techniques depend on this assumption.

Figure 21-5 shows a time series plot of ExxonMobil's daily returns throughout 2013.

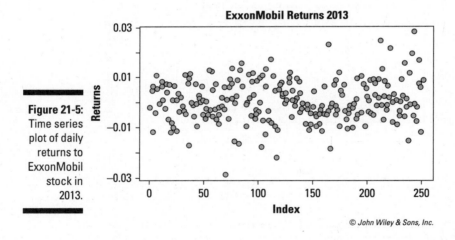

Figure 21-5:
Time series plot of daily returns to ExxonMobil stock in 2013.

The graph shows that as time elapses, the observations seem to be centered around zero. This indicates that the mean isn't changing over time. If the mean were rising over time, the points on the graph would tend to shift up; if the mean were falling over time, the points on the graph would tend to shift down.

Does the Spread of the Dataset Change Over Time?

For time series data, it's also important to know whether the variance of the data is changing over time. Figure 21-5 shows that as time passes, the spread among the observations is growing steadily. (That is, data is becoming more spread out as time elapses.) This indicates that the variance (as well as the standard deviation) is increasing over time.

If the variance is changing over time, that can cause serious problems for many statistical techniques. Fortunately, there are methods available that can correct for this problem.

The situation where variance isn't constant over time has a very intimidating name in econometrics: *heteroscedasticity*. Pronouncing this word isn't easy! Check out www.forvo.com/word/heteroscedasticity/ to hear it spoken.

Are There Outliers in the Data?

An *outlier* is a member of a dataset that is significantly larger or smaller than the other values in the dataset. (See Chapter 10 for a discussion of outliers.) Outliers can greatly affect some statistical tests, so it's important to determine whether outliers are present, and if so, whether they should be removed from the dataset.

An outlier may be defined in terms of quantiles, as follows:

- ✔ If an observation is less than $Q_1 - 1.5(IQR)$, it's considered to be an outlier.
- ✔ If an observation is greater than $Q_3 + 1.5(IQR)$, it's considered to be an outlier.

A box plot shows outliers as individual points at the top and bottom of the plot. Figure 21-2 shows that in the ExxonMobil data, there are three outliers greater than $Q_3 + 1.5(IQR)$ and two outliers smaller than $Q_1 - 1.5(IQR)$. These could have been caused by several factors, such as unexpectedly good or bad data released by the company, surprisingly large changes in oil prices, and so forth.

Does the Data Conform to Our Assumptions?

Many types of statistical analysis depend on key assumptions about the data. Chapter 8 discusses this notion at length. Some of the most commonly used assumptions include the following:

- Normally distributed data
- Independent observations in the data
- Constant parameters (mean, variance, and standard deviation) for the observations in the data
- No outliers in the data

EDA techniques enable you to test these assumptions before proceeding with formal statistical tests. The results for the ExxonMobil data given in this chapter show the following:

- The data is close to being normally distributed.
- The members of the data are very nearly independent of each other.
- The mean appears to be constant over time.
- The variance appears to be increasing over time.
- There are several outliers in the data.

Index

• *R* •

• *S* •

• U •

About the Authors

Alan Anderson, PhD is a professor of economics and finance at numerous schools, including Fordham University and New York University. Outside of the academic environment he has many years of experience working as an economist, risk manager, and fixed income analyst. Alan received his PhD in economics from Fordham University, and an M.S. in financial engineering from Polytechnic University. He is the author of *Business Statistics For Dummies* (Wiley, 2013).

David Semmelroth has spent the last two decades working with data and training people to work with data. He has helped develop increasingly sophisticated techniques to collect, manage, and use data to drive business results. His industry experience includes working with a variety of financial services companies, from insurance to retail banking. More recently he's been doing consulting work related to customer databases and database marketing. He also spent several years in the travel and entertainment industry. David earned both his B.A. and M.S. degrees in mathematics from the University of Michigan and has taught a variety of courses in statistics and mathematics at the college level. He is the author of *Data Driven Marketing For Dummies* (Wiley, 2014).

Authors' Acknowledgments

The authors would like to thank their editors and everybody else at Wiley who helped make this book happen.

Publisher's Acknowledgments

Acquisitions Editor: Lindsay Lefevere

Editor: Corbin Collins

Technical Editor: Karen R. Hum, PhD

Special Help: Barry Schoenborn

Production Editor: Kinson Raja

Cover Image: ©iStock.com/loops7

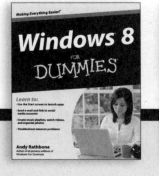

Math & Science

Algebra I For Dummies,
2nd Edition
978-0-470-55964-2

Anatomy and Physiology
For Dummies, 2nd Edition
978-0-470-92326-9

Astronomy For Dummies,
3rd Edition
978-1-118-37697-3

Biology For Dummies,
2nd Edition
978-0-470-59875-7

Chemistry For Dummies,
2nd Edition
978-1-118-00730-3

1001 Algebra II Practice
Problems For Dummies
978-1-118-44662-1

Microsoft Office

Excel 2013 For Dummies
978-1-118-51012-4

Office 2013 All-in-One
For Dummies
978-1-118-51636-2

PowerPoint 2013
For Dummies
978-1-118-50253-2

Word 2013 For Dummies
978-1-118-49123-2

Music

Blues Harmonica
For Dummies
978-1-118-25269-7

Guitar For Dummies,
3rd Edition
978-1-118-11554-1

iPod & iTunes
For Dummies, 10th Edition
978-1-118-50864-0

Programming

Beginning Programming
with C For Dummies
978-1-118-73763-7

Excel VBA Programming
For Dummies, 3rd Edition
978-1-118-49037-2

Java For Dummies,
6th Edition
978-1-118-40780-6

Religion & Inspiration

The Bible For Dummies
978-0-7645-5296-0

Buddhism For Dummies,
2nd Edition
978-1-118-02379-2

Catholicism For Dummies,
2nd Edition
978-1-118-07778-8

Self-Help & Relationships

Beating Sugar Addiction
For Dummies
978-1-118-54645-1

Meditation For Dummies,
3rd Edition
978-1-118-29144-3

Seniors

Laptops For Seniors
For Dummies, 3rd Edition
978-1-118-71105-7

Computers For Seniors
For Dummies, 3rd Edition
978-1-118-11553-4

iPad For Seniors
For Dummies, 6th Edition
978-1-118-72826-0

Social Security
For Dummies
978-1-118-20573-0

Smartphones & Tablets

Android Phones
For Dummies, 2nd Edition
978-1-118-72030-1

Nexus Tablets
For Dummies
978-1-118-77243-0

Samsung Galaxy S 4
For Dummies
978-1-118-64222-1

Samsung Galaxy Tabs
For Dummies
978-1-118-77294-2

Test Prep

ACT For Dummies,
5th Edition
978-1-118-01259-8

ASVAB For Dummies,
3rd Edition
978-0-470-63760-9

GRE For Dummies,
7th Edition
978-0-470-88921-3

Officer Candidate Tests
For Dummies
978-0-470-59876-4

Physician's Assistant Exam
For Dummies
978-1-118-11556-5

Series 7 Exam For Dummi
978-0-470-09932-2

Windows 8

Windows 8.1 All-in-One
For Dummies
978-1-118-82087-2

Windows 8.1 For Dummie
978-1-118-82121-3

Windows 8.1 For Dummie
Book + DVD Bundle
978-1-118-82107-7

Available in print and e-book formats.

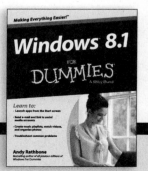

Available wherever books are sold. **For more information or to order direct visit www.dummies.com**

Take Dummies with you everywhere you go!

Whether you are excited about e-books, want more from the web, must have your mobile apps, or are swept up in social media, Dummies makes everything easier.

Leverage the Power

For Dummies is the global leader in the reference category and one of the most trusted and highly regarded brands in the world. No longer just focused on books, customers now have access to the For Dummies content they need in the format they want. Let us help you develop a solution that will fit your brand and help you connect with your customers.

Advertising & Sponsorships

Connect with an engaged audience on a powerful multimedia site, and position your message alongside expert how-to content.

Targeted ads · Video · Email marketing · Microsites · Sweepstakes sponsorship

21 Million Monthly Page Views & 13 Million Unique Visitors